THE STARS OF HEAVEN

♉ Works by Clifford A. Pickover

The Alien IQ Test

Black Holes: A Traveler's Guide

Chaos and Fractals

Chaos in Wonderland

Computers, Pattern, Chaos, and Beauty

Computers and the Imagination

Cryptorunes: Codes and Secret Writing

Dreaming the Future

Future Health: Computers and Medicine in the Twenty-first Century

Fractal Horizons: The Future Use of Fractals

Frontiers of Scientific Visualization (with Stu Tewksbury)

The Girl Who Gave Birth To Rabbits

Keys to Infinity

The Loom of God

Mazes for the Mind: Computers and the Unexpected

Mind-Bending Puzzles (Calendars)

The Paradox of God and the Science of Omniscience

The Pattern Book: Fractals, Art, and Nature

The Science of Aliens

Spider Legs (with Piers Anthony)

Spiral Symmetry (with Istvan Hargittai)

Strange Brains and Genius

Surfing Through Hyperspace

Time: A Traveler's Guide

Visions of the Future

Visualizing Biological Information

Wonders of Numbers

The Zen of Magic Squares, Circles, and Stars

The Stars of Heaven

Clifford A. Pickover

OXFORD
UNIVERSITY PRESS

OXFORD

UNIVERSITY PRESS

Oxford   New York
Auckland   Bangkok   Buenos Aires
Cape Town   Chennai   Dar es Salaam   Delhi   Hong Kong   Istanbul
Karachi   Kolkata   Kuala Lumpur   Madrid   Melbourne   Mexico City   Mumbai
Nairobi   São Paolo   Shanghai   Taipei   Tokyo   Toronto

First published by Oxford University Press, Inc., 2001
First issued as an Oxford University Press paperback, 2004
198 Madison Avenue, New York, New York 10016

www.oup.com

Oxford is a registered trademark of Oxford University Press

Library of Congress Cataloging-in-Publication Data
Pickover, Clifford A.
The stars of heaven / Clifford A. Pickover.
p. cm.
Includes bibliographical references and index.
ISBN 0-19-514874-6 (cloth)
ISBN 0-19-517159-4 (pbk.)
1. Stars. 2. Astrophysics. I. Title.
QB801 .P53 2001
523.8—dc21   2001032870

1 3 5 7 9 8 6 4 2

Printed in the United States of America
on acid free paper

This book is dedicated to the triple alpha process

and the number **7.6549**:

the reasons we

are alive today,

and smiling,

on Earth.

God gave us the darkness so we could see the stars.
— Johnny Cash, "Farmers' Almanac"

O God, guide me, protect me,
make of me a shining lamp and a brilliant star.
— Abdu'l-Bahá

And the third angel sounded,
and there fell a great star from heaven,
burning as it were a lamp.
— Revelation 8:10

Know thou that every fixed star hath its own planets,
and every planet its own creatures,
whose number no man can compute.
— *Baha'u'llah*

Be humble for you are made of dung.
Be noble for you are made of stars.
— Serbian proverb

♑ Acknowledgments

I'll tell you what the Big Bang was, Lestat.
It was when the cells of God began to divide.

— Anne Rice, *Tale of the Body Thief*

I owe a special debt of gratitude to physicist and astronomer Dr. Dina Moché, for her wonderful books and papers from which I have drawn many facts regarding the stars. I heartily recommend her book *Astronomy* for a lively, up-to-date account of the wonders of planets, stars, and galaxies. I also thank Marcus Chown, author of *The Magic Furnace,* a compelling story of stellar nucleosynthesis and the "miracle" of the triple alpha process. Kirk Jensen, Clay Fried, Dennis Gordon, and David Glass have made valuable suggestions regarding the manuscript.

Finally, I thank Samuel Marcius for symbols such as and , which represent Miss Muxdröözol and Mr. Plex.

♌ Contents

Introduction   x

☆ CHAPTER 1
Stellar Parallax and the Quest for Transcendence   1

☆ CHAPTER 2
The Joy and Paschen of Starlight   14

☆ CHAPTER 3
Spectral Classes, Temperatures, and Doppler Shifts   39

☆ CHAPTER 4
Luminosity and the Distance Modulus   58

☆ CHAPTER 5
Hertzsprung-Russell, Mass-Luminosity Relations,   71
and Binary Stars

☆ CHAPTER 6
Last Tango on the Heliopause   90

☆ CHAPTER 7
Stellar Evolution and the Helium Flash   117

☆ CHAPTER 8
Stellar Graveyards, Nucleosynthesis, and Why We Exist   142

☆ CHAPTER 9
Some Final Thoughts   188

Notes   198

Appendix 1. Stars in the Bible   212

Appendix 2. Updates and Breakthroughs   217

Further Reading   224

About the Author   226

Index   229

♉ Introduction

I like the stars. It's the illusion of permanence, I think. I mean, they're always flaring up and caving in and going out. But from here, I can pretend . . . I can pretend that things last. I can pretend that lives last longer than moments. Gods come, and gods go. Mortals flicker and flash and fade. Worlds don't last; and stars and galaxies are transient, fleeting things that twinkle like fireflies and vanish into cold and dust. But I can pretend.

— Neil Gaiman, *The Sandman #48: Journey's End*

Smilodon Overdrive

Unknowingly, we plow the dust of stars, blown about us by the wind, and drink the universe in a glass of rain.

— Ihab Hassan, *The Right Promethean Fire*

The other day I was walking in a field when I came upon a large skull. It was probably from a bear, although I like to imagine it was part of the remains of a prehistoric mammal that once roamed Westchester County, New York. I'm a collector of prehistoric skulls. In my office, I have a skull of a saber-tooth tiger, also known to scientists as the *smilodon*. This killing machine had huge, dagger-like canine teeth and a mouth that could open 90 degrees to clear the sabers for their killing bite.

When I run my fingers lingeringly over the skulls, I am sometimes reminded of stars in the heavens. Without stars, there could be no skulls. The elements in bone, like calcium, were first created in stars and then blown into space when the stars died (figure I.1). Without stars there would be no elements heavier than hydrogen and helium, and, therefore, life would never have evolved. There would be no planets, no microbes, no plants, no tigers, no humans.

Now I look at the saber-tooth tiger's skull, so massive, so deadly. Without stars, the tiger racing across the savanna fades away, ghostlike. There are no iron atoms for its blood, no oxygen for it to breathe, no carbon for its proteins and DNA. There are no mossy caverns, mist-covered swamps, black vipers, retinas, spiral nautilus shells. Our existence requires stars to forge the heavy ele-

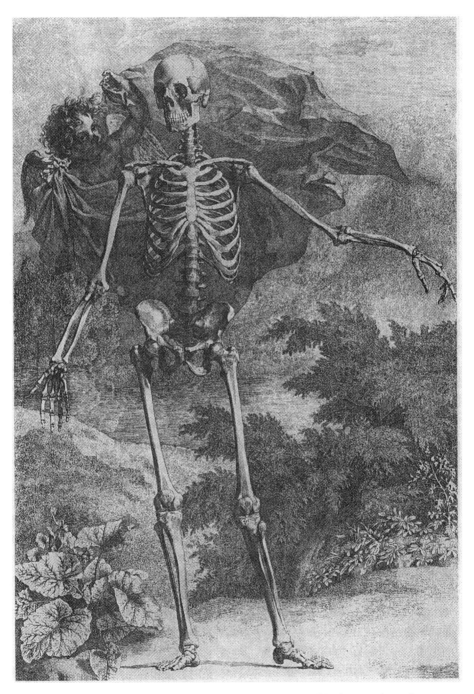

Figure 1.1   The calcium in our bones was first created in furnaces located
in the center of stars. [From Robert Beverly Hale and Terence Coyle,
Albinus on Anatomy (New York: Dover, 1979), 29.]

ments in massive fusion reactions, but we also need the stars to explode at the ends of their lives to wash the new elements far into space. Without these supernova explosions, there are no seagull cries, computer chips, trilobites, Beethovens, or the tears of a little girl. There is no Golgotha, and Jesus never gave his Sermon on the Mount. There is no one to speak the words, "Thy will be done on Earth, as it is in heaven." Without exploding stars, perhaps there would be a heaven, but there certainly would be no Earth.

Let's imagine the origin of calcium in this tiger tooth that I hold. The atoms created in the dying ancient stars were blown across vast distances and eventually formed the elements in the planets that coalesced around our Sun. If you could turn back time and follow the carbon atoms in the tiger's brain back to their source, you would connect the tiger to an unimaginably long interstellar journey that culminated in the giant stars, which died in violence billions of years ago. Humans and tigers and whales and plants and all that we see on Earth are stellar ashes. And when the ancient tiger died, the atoms in its flesh kept going. Perhaps one of the tiger's atoms coalesced into your embryonic form.

I know I almost sound religious when I tell my friends about the stars. We are lucky we live in an age in which we can wonder about the myriad cosmic "coincidences" that permitted the creation of stars and the flushing of their heavy elements to the Universe. I'll explain in this book how various nuclear and chemical constants are precariously poised to permit life. It is as if the constants sat on the head of a pin, the tiny point of which encourages a Universe full of complex compounds rather than the seemingly more likely oceans of monotonous hydrogen. We are the memorial to shattered stars. We are the afterlife of which blazing stars could never dream.

The elements in our world are constantly changing. With every breath, we inhale millions of atoms of air exhaled a few days ago by someone on the other side of the planet. In some sense, our brains and organs are vanishing into thin air, the cells being replaced as quickly as they are destroyed. The entire skin replaces itself every month. Our stomach linings replace themselves every five days. We are always in flux. A year from now, 98 percent of the atoms in our bodies will have been replaced with new ones. We are nothing more than a seething mass of moving atoms, continuous threads in the fabric of spacetime. What does it mean that your body has nothing in common with the body you had a few years ago? If you are something other than the collection of atoms making up your body, what are you? You are not so much your atoms as you are the *pattern* in which your atoms are arranged. Very likely you have the atoms of the great Biblical prophets coursing through your body, and they, like all of us, breathed in the atoms from the stuff of stars. As mathematician Rudy Rucker has noted, "The simple processes of eating and breathing weave all of us together into a vast four-dimensional array. No matter how isolated you may sometimes feel, no matter how lonely, you are never really cut off from the whole."[1]

Who are we? Where do we come from? In Joni Mitchell's 1960s song *Woodstock* we hear the answer, "We are stardust, billion year old carbon."[2] In this book, I will help you to better understand the meaning of these words.

Starry Night

The heavens call to you, and circle about you, displaying to you their eternal splendors, and your eye gazes only to earth.

— Dante Alighieri, *Purgatorio*

Stars have fascinated us since the dawn of history and have allowed us to transcend ordinary lives in both literature and the arts. I think the painter Vincent van Gogh glimpsed a poignant portion of reality at the height of his interest in stars. van Gogh loved to read about astronomy and wondered about what it would mean to travel to the stars. His famous *Starry Night*, painted in 1889, shows stars not as points of lights but as bright orbs with the sky swirling about them like a magical stream (figure I.2). Van Gogh contemplated new ways of painting stars so they revealed their glory and took over the canvas. He also thought about what stars might mean to humans and their place in the Universe.

Around the time van Gogh painted *Starry Night*, he wrote about stars to his brother Theo. His letter seems to be a meditation on the realm between life and death and perhaps how death might be a portal to the stars:

Is the whole of life visible to us, or do we in fact know only the one hemisphere before we die? For my part I know nothing with any certainty, but the sight of the stars makes me dream, in the same simple way as I dream about the black dots representing towns and villages on a map. Why, I ask myself, should the shining dots in the sky be any less accessible to us than the black dots on the map of France? If we take the train to get to Tarascon or Rouen, then we take death to go to a star. What is certainly true in this reasoning is that while we are alive we cannot go to a star, any more than, once dead, we could catch a train. It seems not impossible to me that cholera, gravel, phthisis [a wasting disease, especially tuberculosis], and cancer could be the means of celestial transportation, just as steam-boats, omnibuses, and railways serve that function on earth. To die peacefully of old age would be to go there on foot.[3]

Figure I.2   *Starry Night* by Vincent van Gogh, June 1889.
[Oil on canvas. 73.7 × 92.1 cm (29 × 36¼ inches).
Collection, The Museum of Modern Art, New York.
Acquired through the Lillie P. Bliss Bequest.
© 2001 The Museum of Modern Art, New York.]

Many authors have speculated that van Gogh had temporal lobe epilepsy and that this brain disorder intensified his religious needs and experiences.[4] For van Gogh, abnormal electrical activity in the brain was a portal that opened doors to entirely new ways of seeing and feeling. He once wrote, "I often feel a terrible need of—shall I say the word?—of religion. Then I go out at night to paint the stars."[5] At the time that he painted *Starry Night* he was in the asylum and had about a year to live. But his mental instability and acuteness did not mean that *Starry Night* was the wild raving of a lunatic or conjured up without observing the sky. He wrote to his brother Theo, "This morning I saw the country from my window a long time before sunrise, with nothing but the morning star, which looked very big." The morning star is another name for Venus, which is probably portrayed as the large bright shimmering form, just to the left of center in his painting. According to University of California–Los Angeles art historian Albert Boime, astronomical data proves the placement of stars and moon in *Starry Night* are accurate for the night on which it is known to have been painted. In particular, Boime has reconstructed the probable alignment of stars and planets in the painting, seeing in the painting three stars of the constellation Aries as well the Moon and Venus.[6]

In some ways van Gogh made us see the Universe in a different light and, with just a few strokes of a paint brush, allowed us to appreciate the vastness of the night sky as much as modern telescope images do. Look at figure I.2. Look at the contrast between the intense turbulence of the heavens and the calm order of the village and church below. The contrast makes the sky resonate in the mind long after your eyes leave the painting.

van Gogh's art is just one example of humanity's passion for stars. In fact, humans have always looked to the stars as a source of inspiration and transcendence to lift them beyond the boundaries of ordinary intuition. The ancient Sumerians, Egyptians, Chinese, and Mexicans were very aware of the locations and motions of the visible stars. Some of these cultures had catalogued and grouped thousands of stars and perhaps thought that the *visible* stars were all the stars that existed.[7] On the other hand, the Old Testament writers theorized there were many more stars than humans could see. According to Genesis 22:17, the stars were as great in number as the sands of the seashore and simply could not be numbered. The vast reaches of the cosmos were utterly incomprehensible to humans: "For as the heavens are higher than the earth, so are my ways higher than your ways, and my thoughts than your thoughts" (Isaiah 55:9).

In the Bible, stars are a sign of God's power and majesty. In Job 38:31–32, God reminds Job of His omnipotence and names several constellations of stars: "Can you bind the beautiful Pleiades?[8] Can you loose the cords of Orion?[9] Can you bring forth the constellations in their seasons or lead out the Bear with its cubs?[10] Do you know the laws of the heavens?" In Isaiah 40:26, God similarly reminds us that He knows their number and their names: "Lift your eyes and look to the heavens: Who created all these? He who brings out the starry host one by one, and calls them each by name. Because of His great power and mighty strength, not one of them is missing."

Probably the most famous star in the Bible occurs in Matthew 2:1–2, which describes a group of travelers, called Magi, heading toward Bethlehem from somewhere in the east. These Magi are most likely astrologers. They had seen a special star and were bringing gifts for "the one who has been born king of the Jews."

> Where is he that is born King of the Jews? For we have seen his star in the east, and are come to worship him. . . . And lo, the star, which they saw in the east,

went before them, till it came and stood over where the young child was. When they saw the star, they rejoiced with exceedingly great joy. (Matthew 2:1–2, 9–10)

Over the centuries numerous scholars have sought a scientific explanation for the Star of Bethlehem (figure I.3). Jesus seems to have been born sometime between 4 and 8 B.C. Chinese annals record novae (bright stars) in 5 B.C. and 4 B.C. In the early seventeenth century Johannes Kepler suggested that the Star of Bethlehem may have been a nova in the constellation of Pisces the Fish occurring near a conjunction of Jupiter and Saturn around 7 B.C.[11] (Coincidentally, a fish has long been a symbol of the Christian church.)

Figure I.3   Wise men guided by the Star of Bethlehem. [From Gustave Dore, *The Dore Bible Illustrations* (New York: Dover Publications, 1974), 163.]

In Islamic theology, the number of stars is also a metaphor for a huge number. In pre-Islamic times, Ka'b Al-Ahbar was one of the great Jewish scholars. He later became a Muslim and said, "On the 15th of [the month of] Shaban, Allah ordered that Paradise be decorated, and then Allah freed from hellfire as many persons as the number of stars in the universe."[12]

Today we can get a feel for the actual number of stars that exist in the Universe. It is clear that there are a lot more stars than contemplated by many early astronomers. For example, well before Italian astronomer Galileo Galilei (1564–1642) developed the first telescopes for astronomical observation in 1609, Greek astronomer Hipparchus (ca. 127 B.C.) compiled a star catalogue containing 850 stars, and Alexandrian astronomer Ptolemy (127–151) increased the number to 1,022 stars. Ptolemy's star catalogue is the oldest surviving star catalogue, and it grouped stars in constellations with the latitude, longitude, and the apparent brightness of each star. Danish astronomer Tycho Brahe (1546–1601) listed accurate positions of more than 777 stars. German astronomer Johannes Kepler (1571–1630) catalogued 1,005 stars. Many later scientists such as Galileo believed that the stars could not be numbered, and the Bible similarly states, "... the host of heaven cannot be numbered."[13]

We live in a gravitational "froth" where gravity binds stars together to form galaxies, binds galaxies into local groups of galaxies, groups of galaxies into clusters, clusters into superclusters, and superclusters into "walls." Luckily for us, the galaxies, with their strong gravitational attraction, consolidate the chemically enriched gas left over from stellar explosions. The number of stars in the Universe boggles the mind. The *variety* is equally amazing—black holes, red giants, brown dwarfs, white dwarfs, Cepheid variables, neutron stars, pulsars ... Our modern, sophisticated telescopes have only begun to reveal the immense numbers and variety of stars. We find that our own Sun is just an ordinary star that inhabits our Galaxy, the Milky Way, which has roughly 200 billion stars. Some of the stars are much bigger—giants and even supergiants. There are around 100 billion galaxies in the observable Universe and each one of them has roughly 100 billion stars. So there is roughly one galaxy for every star in the Milky Way.[14]

When students ask astronomer William Keel of The University of Alabama in Tuscaloosa how many stars exist in our Milky Way Galaxy, his standard answer is "about as many as the number of hamburgers sold by McDonald's." It is difficult to be precise because distance and dust absorption dim incoming light. Measurements of the relative numbers of stars with different absolute brightness suggests that for every Sun-like star there are about 200 faint red M-class dwarfs. (As you'll learn, the "class" of a star is determined by its surface temperature. M stars are cool.) This means that to estimate the number of stars in the Milky Way, we must consider the number of luminous stars that we can see at large distances and assume that we know the proportion of visible stars to the invisible fainter stars.[15]

Incidentally, the diameter of the Universe we see right now—we call it the *observable Universe*—is about 10^{26} meters, which is 1 followed by 26 zeros. You might enjoy comparing this distance to a few other distances for comparison (Table I.1):

Table I.1
Distances[16]

Diameter (meters)	Object
1.3×10^7	Earth (diameter at equator)
1.5×10^7	Sirius B, a white dwarf star
1.4×10^8	Jupiter
7.7×10^8	Moon's orbit
1.4×10^9	Sun
4×10^{10}	Rigel, a blue-white giant
3×10^{11}	Earth's orbit (average diameter)
1.2×10^{12}	Betelgeuse, a red supergiant
1.5×10^{13}	Solar system
2×10^{14}	Heliopause (edge of solar wind)
10^{15}	Bok globule (a nebula from which a star is formed)
9.5×10^{15}	Light-year
10^{16}	Planetary nebula (formed by outgassing from a star)
4×10^{16}	Distance to closest star, Proxima Centauri
1.6×10^{18}	M13, typical globular cluster
9×10^{20}	Milky Way's disc
6×10^{22}	Local Group (a cluster of around 30 galaxies containing the Milky Way)
2×10^{23}	Typical cluster (containing 100–1,000 galaxies)
2×10^{24}	Typical supercluster (containing 3–10 clusters)
10^{26}	Observable universe

Current theories of the early Universe suggest it inflated faster than the speed of light; therefore, we will never see some of the very distant parts of the Universe.[17] This means that the observable Universe is only that part that is accessible to us, given the age of the Universe and the finite speed of light. If the inflationary theory is correct, then even with the most amazing telescopes we will ever develop, we will only see a very small part of all that exists.

The Science and Spirituality of Stars

Why should the universe be constructed in such a way that atoms acquire the ability to be curious about themselves?

— Marcus Chown, *The Magic Furnace*

This book will allow you to travel through time and space, and you needn't be an expert in astronomy or physics. To facilitate your journey, I start most chapters with a dialogue between two or three quirky explorers who study stars. Bob is chief curator of an intergalactic art museum, a teacher, and a star enthusiast. His able student is a scolex, a member of a race of creatures with strong diamond bodies. His personal scolex, Mr. Plex, will do *whatever* experiments Bob wishes.[18]

Prepare yourself for a strange journey as *The Stars of Heaven* unlocks the doors of your imagination with thought-provoking mysteries, puzzles, and problems on topics ranging from stellar anatomy to the birth of solar systems. A resource for science-fiction writers, an adventure and education for beginning physics and astronomy students, each chapter is a world of paradox and mystery.

Imagery is at the heart of much of the work described in this book. To understand what is around us, we need eyes to see it. I hope the numerous diagrams help convey the concepts from myriad perspectives. As in all my previous books, you are encouraged to pick and choose from the smorgasbord of topics. Many of the chapters are brief and give you just a flavor of an application or method. Additional information can be found in the referenced publications. Some information is repeated so that each chapter contains sufficient background information, but I suggest you read the chapters in order as Bob and Mr. Plex gradually build their knowledge. By the time you finish this book, you'll be able to impress your friends with such arcane phrases as the Rydberg-Ritz formula, Paschen series, heliopause, helium flash, triple alpha processes, and Hertzsprung-Russell.

In about five billion years, the hydrogen fuel in our Sun will be exhausted in its core, and the Sun will begin to die and dramatically expand, becoming a red giant. At some point, our oceans will boil away. No one on Earth will be alive to see a red glow filling most of the sky. As Freeman Dyson once said, "No matter how deep we burrow into the Earth . . . we can only postpone by a few million years our miserable end."

Where will humans be, a few billion years from now, at the End of the World?[19] Even if we could somehow withstand the incredible heat of the Sun, we would not survive. In about seven billion years, the Sun's outer "atmosphere" may engulf the Earth. Due to atmospheric friction, in many scenarios the Earth will spiral into the sun and incinerate. However, I don't think we have to mourn for humanity. In five billion years, humans will probably have downloaded their minds to computers, left the solar system in some great diaspora, and sought their salvation in the stars.

> Rationalists predicted that religion would be the first thing to fall when humanity went to the stars and found no gods. . . . Scientists never had been all that good at predicting . . . They never even noticed that, when they finally went out there, every deity and supernatural belief system known at the time went right with them . . .
>
> — Jack Chalker, *Balshazzar's Serpent*

> And when he opened the seventh seal, there was silence in heaven about the space of half an hour.
>
> — Book of Revelations, 8:1

CHAPTER 1

Stellar Parallax and the Quest for Transcendence

Come quickly, I am tasting stars!

— Dom Pérignon (1638–1715),
    at the moment of his discovery of champagne

The year is 2100, and Bob is chief curator of an intergalactic art museum. Nicknamed "Picasso," his large ship has artworks from several star systems. Bob is currently hovering above the Earth, and on his view-screen is a nearby star.

Several cleaning spiders work their way along the plush carpeting of Bob's living quarters. They occasionally ingest drops of spilled paint, lint, candy wrappers, and other detritus about which Bob is better off not thinking. It never ceases to amaze Bob how robotics have became the Solar System's largest industry, eclipsing the information industry. The latter became important by automating office work, bookkeeping, communications, and calculations. Robotics automated everything else.

Bob turns to his assistant. "Mr. Plex, that star is Sirius in the constellation Canis Major." Bob pronounces the words "SEAR-ee-us" and "KAY-niss MAY-jer."

"Very beautiful, Sir. And it's a bright one."

Bob nods. "It's the brightest star in the night sky. If we magnify the image you can see that it's actually a binary star, two stars with the brighter one 23 times as bright as our own Sun. Today I want to teach you about how earlier astronomers determined the distances from stars to Earth."

Bob's dual faces smile. Stars are his life.

Mr. Plex looks around Bob's office, slowly gazing at photographs and paintings of constellations, galaxies, and stars in various stages of their evolution. He rests his gaze on Bob's favorite work by an extraterrestrial artist named Miss Muxdröözol. Muxdröözol, a trochophore with two huge teardrop shaped eyes, has painted an exploding star using the blood of several ancestors.

Bob sometimes wonders about Miss Muxdröözol. Her skin is exceptionally smooth but her shape is somewhat disconcerting. She essentially has no body— just a large head connected to arms and legs.

1

Miss Muxdröözol

Bob shakes his head. "Mr. Plex, for the next few days I will teach you every-thing I know about the wonderful stars in our Universe. Today I want to start by having you help determine the distance of Sirius from Earth using *stellar paral-lax*, a simple, elegant method for understanding our place in the Universe."

"Sounds delightful. Shall I get Miss Muxdröözol?"

Bob looks at Mr. Plex with one face and then the other. "No, let her sleep."

Mr. Plex stares for a second at Bob's primary face. It is the clean-shaven face of a man about 35 years old. Bob's eyes are brown and his curly hair shows signs of premature gray. Bob's secondary face is similar except that one of the eyes is blue.

Yes, Bob has two faces, 180 degrees opposed from one another. Is he a set of Siamese twins? A product of genetic engineering? A freak of nature? Bob always says, "If you have to ask, I don't want to know you." Bob is delighted that his name is a palindrome and feels that the test of a wise people is their ability to hold two opposed ideas in their minds at the same time and still remain sane.

Mr. Plex, Bob's first officer, is a scolex, a member of a race of creatures with diamond reinforced exoskeletons that allow them to explore outer space with little consequences to their health. Protruding from the bottom of Mr. Plex's huge head and gaping mouth are four appendages that serve as both feet and arms.

Mr. Plex

"Mr. Plex, I want you to leave our ship and propel yourself towards the other side of the Earth. When you get there, transmit an image of the star back to me."

"Yes sir," Mr. Plex says. His slight hesitation is betrayed by the slight clatter-ing of his canines and the quivering of his anterior forelimb.

"Sir, how will this help?"

Bob's voice is firm. "I will explain when you come back."

Bob watches his scolex leave the ship. While waiting, Bob wanders through some of the nearby rooms of the art museum, rooms devoted to Egyptology, Greco-Roman archeology, Byzantine art, Chinese artifacts, and the hall of Smilodon and Troödon skulls. The Troödon was probably the smartest of the dinosaurs, with a relatively large brain for its body size.

An hour later Bob is back in his office, just in time to hear Mr. Plex say, "I am now staring at Sirius from the other side of Earth. I just captured an image and stored it in my detachable brimp." (Scolexes frequently use brimps, short for brain implants, to store information they cannot possibly retain in their natural brains.) (Figure 1.1)

After some time, Bob's scolex returns to the ship. His breathing is rapid and heavy as he makes an oboe-like sound. Bob smells the faint odor of a scent resembling musk.

Bob waits a few moments for him to catch his breath. "Mr. Plex, let's look at your two star images—the one we took from the ship and the one you took from the other side of the Earth."

"They look identical to me, Sir."

"That's right. The distance you traveled was not enough to demonstrate stellar parallax. The best way to make you remember this was to send you out and have you return to the ship. I'm sure you will never forget this lesson."

Mr. Plex's lung lobes quiver like leaves in a storm.

"However, parallax could be used to measure the distance to Sirius in the following way. First we would determine the position of Sirius relative to other stars. Half a year later, when the Earth has traveled halfway around the Sun, we

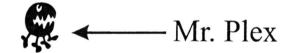 Mr. Plex

 Earth

 Sirius

Figure 1.1   Mr. Plex looking at the star Sirius from two different positions
(diagram not to scale.)

would measure Sirius's position again." Bob sketches a figure on the flexscreen video wall (figure 1.2).

Mr. Plex nods as he touches Bob's diagram. "Yes, it looks like Sirius moves against the backdrop of more distant stars depending on where we are when we measure the location from Earth."

"Exactly. That's what stellar parallax is all about. Stellar parallax is similar to effects we can see when closing one of our eyes. Look at my hand with one eye, then the other. My hand seems to move. Similarly, Sirius seems to change locations when we change our observation point, and we can calculate the distance to Sirius from its *parallax angle*." Bob writes "parallax angle" on his drawing (figure 1.2). "Of course, these angles are tiny, but I've drawn them large to help illustrate the concept."

"Tiny? That's why I couldn't see a difference in the angle when I simply traveled around the Earth. Your human astronomers in the past had to wait for the Earth to move a large distance[1] before remeasuring the star's location."

"Exactly right. The parallax angles are so tiny that they are measured in seconds of arc where one second is 1/3600 of a degree. If you could view a ladybug, carpenter ant, or a circular piece of paper made from a hole puncher from a mile (or two kilometers) away, they would appear to have a diameter corre-

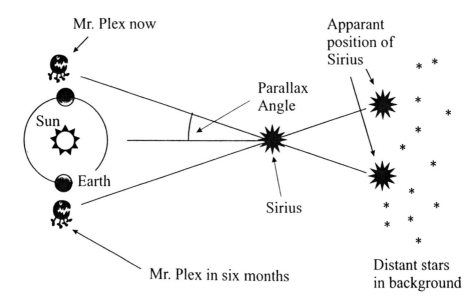

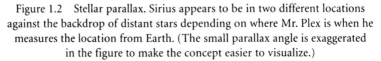

Figure 1.2   Stellar parallax. Sirius appears to be in two different locations against the backdrop of distant stars depending on where Mr. Plex is when he measures the location from Earth. (The small parallax angle is exaggerated in the figure to make the concept easier to visualize.)

sponding to one second! In fact, all of the Earth's nearby stars have a stellar parallax of less than one second."

"Sir, what's an ant?"

Bob reaches into his pocket and withdraws a small creature that looks like:

Bob hands the ant to Mr. Plex and then sketches a table on the flexscreen:

	Table 1.1	
Star	Distance From Earth (light-years)	Parallax (seconds)
Alpha Centauri[2]	4.3	0.75
Barnard's Star	6	0.545
Wolf 359	7.7	0.421

"A light year is a big distance—the distance that light travels in one year. (Light travels 300 thousand kilometers per second or 186 thousand miles per second.) There are 3.26 light years per parsec."

"Sir, I'm dying to see a simple formula that allowed your ancient astronomers to calculate how far away a star would be given its parallax."

Bob grins. "Thank you for asking. One parsec is the distance from a star to Earth for a star that has a parallax of one second. One parsec is 31 trillion kilometers or 19 trillion miles. Here's the formula." Bob hands Mr. Plex a card with the following symbols:[3]

$$\delta = \frac{1}{P}$$

"Here the Greek letter delta is the star's distance from Earth in parsecs, and P is the parallax in seconds of arc. You can see that as the distance increases the parallax decreases. This means that the ancient astronomers could only use this for stars that weren't too far away. Otherwise the parallax would be too small to measure. Our early astronomers could measure distances to about 100 parsecs using this method. Way back in the early 1990s, the European High Precision Parallax Collecting Satellite, or Hipparcos, measured the parallaxes of over 100,000 stars."[4]

Mr. Plex nods and drops the ant. A cleaning spider, the size of a human hand, makes a twittering sound and starts toward the fallen ant. Bob points a finger at the spider and shouts, *"Hör auf zu reden!"* The spider freezes and retreats to an orifice in the floor. Bob picks up his ant.

Bob yawns. "Let's go for a walk as we finish our discussion." They take the elevator to the eighth floor, which is kept at a constant 90 percent humidity to make some of the alien artists feel at home. The air smells of machine oil and wine.

Mr. Plex looks up. The light is dim orange from the overhead receptacles of bioluminescent insects. Mostly it's quiet, except for the throbbing of the life-support machines, the hum of various air pumps, and the clacking of Mr. Plex's five metatarsal nails against the tile floor.

On one of the walls is a calcite plaque dedicated to all the artists who have died while creating great works of art in poisonous or extreme ecosystems. The inscription reads, RMᛟᚲᛈ ᛁᚾ ᚲMᛈᚲM, the runic equivalent of Rest In Peace. Embedded at the top of the plaque is a detached vocal apparatus from a deceased Mangoid. Every few seconds an electrical current innervates a cranial nerve that causes the organ to emit a soft wailing sound.

Bob looks away from the wall and back to Mr. Plex. "Let's calculate the distance to Sirius. The parallax is 0.38 seconds. Using the formula, this means that Sirius is 1/.38 or 2.6 parsecs from Earth. This also means that Sirius is 8.5 light years away or 50 trillion miles away."

A twin shiver goes up Bob's bifurcated spine. "I am always awed by the fact that early astronomers figured out a way to compute a 50-trillion-mile distance without leaving the confines of their backyards. Now that's a real testament to human ingenuity. It's as if humans were measuring the heavens with something the size of ants. Stellar parallax connects the huge to the tiny, and the smallest of parallax angle implies distances we can barely comprehend with our limited brains."

Bob and Mr. Plex walk in silence past the east wall of the museum's main corridor, which leads to seventeen rooms, each of which has been decorated by pseudomorphs, intelligent, fur-covered aliens from planets near the Magellanic Clouds. The mosaics on the floors and ceilings depict Marc Chagall's impressionistic works. To the right is Chagall's "Self-Portrait with Seven Fingers" and "I and the Village." To the left is a holoscreen of "Hommage à Apollinaire" and "Paris Through the Window," which seems to give the artworks the illusion of depth. Some of the art gallery rooms are perfect spheres while others change dimensions constantly. One room is three stories tall, with the lower half walled with sapphires.

The museum is closed at this late hour. Most of the curators have gone to sleep. The only figure that moves is a Ninoan who has evolved from a race of octopoid beings. She is creating a lively, realistic painting on the museum wall.

It shows an aquamarine palace in front of a leaping goose-like figure. Next to the Ninoan is a creature with translucent skin quietly watching, her hair a waving maze of gossamer threads. The collagen and lymph that has once filled the Ninoan's throat appendages trickles to the floor like raindrops in a spring shower. Is she dying, or is the artist making some deep commentary on life? Is this perhaps her species way of expressing emotion? Bob is not sure.

Along the floor, a virtual zoo of robotic devices parade from all over the cosmos. Bob marvels at how the tiny machines with nervous systems modeled after ants display spontaneous social behavior.

Except for the gentle hum of Mr. Plex's breathing holes, the rest of the museum is silent. On the floor, Bob's ants have left a pithy message:

STELLAR PARALLAX

Some Science Behind the Science Fiction

I believe a leaf of grass is no less than the journeywork of the stars"

— Walt Whitman, "Leaves of Grass"

Our quest for determining the distance of stars from Earth has had a long history. The Greek philosopher Aristotle (384–322 B.C.) and Polish astronomer Copernicus (1473–1543) knew that if the Earth orbited our Sun, one would expect the stars to apparently shift back and forth each year. Unfortunately, Aristotle and Copernicus never observed the tiny parallaxes involved, and humans had to wait until the nineteenth century before parallaxes were actually discovered. We'll get to that in a moment, but it is interesting to note that stellar parallax had an impact on the Catholic Church's battle with Galileo because of his assertion that the Earth moves around the Sun (figure 1.3). During his debates with this Church, Galileo never really had proof of this heliocentricism that was sufficiently direct enough to convince the Church. Admittedly, scientific evidence may never convince a faith-based institution of anything, but Galileo could not even answer Aristotle's strongest argument against it:

If there is no stellar parallax, the Earth does not orbit the Sun.

In other words, there should be a shift in the apparent position of a star observed from the Earth when the Earth is on one side of the Sun and then six months

Figure 1.3   Galileo Galilei (1564–1642). Would the Church have reprimanded Galileo if he was able to demonstrate stellar parallax? (Drawing by K. Llewellyn Blakeslee.)

later from the other side. Even with his telescopes, Galileo was not able to observe the slightest stellar parallax. However, Galileo's discovery of Jupiter's moons did dispel the notion that the Earth is the center of all astronomical motion.

In 1614, Galileo told Church officials that heliocentrism contradicted the Bible. For example, Galileo explained that the apparent revolution of the Sun around the Earth refutes the biblical story that Joshua made the *Sun* stand still by appealing to God. (On the other hand, if Joshua actually halted the Earth's spinning on its axis, there could be numerous catastrophes.) Other Bible passages, if taken literally, also imply a static Earth. For example, in Psalms 92, we find, "He has made the world firm, not to be moved." In 1633, the Holy Office finally condemned Galileo as "vehemently suspected of heresy." He was sentenced to life imprisonment, and told to renounce heliocentrism and never to speak of it again. (Galileo was never in a jail or tortured; he stayed mostly at the house of a Vatican ambassador.) In particular Galileo was

> commanded and enjoined, in the name of His Holiness the Pope and the whole Congregation of the Holy Office, to relinquish altogether the said opinion that the Sun is the center of the world and immovable and that the Earth moves; nor further to hold, teach, or defend it in any way whatsoever, verbally or in writing; otherwise proceedings would be taken against him by the Holy Office.[5]

Let's take a few steps back in time. Until his encounters with the Church, Galileo did not usually allow his religious faith and his study of science to interfere with each other. Before he was reprimanded, he wrote on December 13, 1613:

> The Bible, although dictated by the Holy Spirit, admits in many passages of an interpretation other than the literal one. And, moreover, we cannot maintain with certainty that all interpreters are inspired by God. Therefore, I think it would be the better part of wisdom not to allow any one to apply passages of Scripture in such a way as to force them to support as true any conclusions concerning nature, the contrary of which may afterwards be revealed by the evidence of our senses, or by actual demonstration. Who will set bounds to human understanding? Who can assure us that everything that can be known in the world is known already? . . . I am inclined to think that Holy Scripture is intended to convince people of those truths which are necessary for their salvation, and which being far above human understanding cannot be made credible by any learning, or by any other means than revelation.[6]

The first strong clerical attack against Galileo took place in December 1614. Father Thomas Caccini of Florence, Italy denounced mathematics as inconsistent with the Bible and detrimental to the State. He used the Joshua passage in the Bible to make this clear (figure 1.4):

Figure 1.4 Joshua commanding the sun to stand still. [From Gustave Dore, *The Dore Bible Illustrations*. (New York: Dover Publications, 1974), 53.]

Then spake Joshua to the LORD in the day when the LORD delivered up the Amorites before the children of Israel, and he said in the sight of Israel, Sun, stand thou still upon Gibeon; and thou, Moon, in the valley of Ajalon. And the sun stood still, and the moon stayed, until the people had avenged themselves upon their enemies. Is not this written in the book of Jasher? So the sun stood still in the midst of heaven, and hasted not to go down about a whole day. And there was no day like that before it or after it, that the LORD hearkened unto the voice of a man: for the LORD fought for Israel.[7]

Galileo replied that science did not have to conflict with the Bible:

I do not think it necessary to believe that the same God who gave us our senses, our speech, our intellect, would have put aside the use of these, to teach us instead such things as with their help we could find out for ourselves, particularly in the case of these sciences of which there is not the smallest mention in the Scriptures; and, above all, in astronomy, of which so little notice is taken that the names of none of the planets are mentioned. Surely if the intention of the sacred scribes had been to teach the people astronomy, they would not have passed over the subject so completely.[8]

On April 12, 1615, a Roman prelate, Robert Cardinal Bellarmine, who also served as a member of the Inquisition and was known as the "Master of Controversial Questions," summarized the Church's view by referring to the Bible:

The words "the sun also riseth and the sun goeth down, and hasteneth to the place where he ariseth, etc." were those of Solomon, who not only spoke by divine inspiration but was a man wise above all others and most learned in human sciences and in the knowledge of all created things, and his wisdom was from God.

One wonders how the Church and the Inquisition would have reacted if Galileo were to find evidence of stellar parallax. Would they have shrugged it off as easily as they did other scientific evidence or would this additional piece of evidence be so strong as to cause them to transcend their literal interpretation of the Bible?

It was not until 1838 that the first stellar parallax was measured. Using a telescope, German astronomer Freidrich Wilhelm Bessel (1784–1846) was studying the star 61 Cygni (SIG-nye) in the constellation of Cygnus the Swan (figures 1.5 and 1.6). 61 Cygni displayed significant apparent motion each year, and Bessel's parallax calculations indicated that the star was 10.3 light-years (3.4 parsecs) away from Earth. Bessel later measured positions for thousands of stars, and he was the first person to accurately determine interstellar distances.

Just visible to the naked eye, we know today that 61 Cygni is actually an attractive binary of two dwarf orange stars, with one star completely orbiting

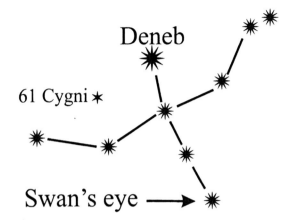

Figure 1.5   The holy grail of Galileo: 61 Cygni of constellation Cygnus, the Swan. Would knowledge of 61 Cygni have caused Galileo's dreaded Inquisitors to transcend their literal interpretation of the Bible?

the other every 653 years. The constellation of Cygnus contains many wonders, including Cygnus X-1, one of the strongest X-ray sources in the sky and possibly a black hole. The brightest star in Cygnus is Deneb (DEN-ebb, or Alpha Cygni; see figure 1.5), which means "the hen's tail" suggesting its position in the tail of Cygnus the Swan. Deneb is approximately 1600 light-years away from Earth, 60,000 times more luminous than our Sun, and 25 times the size of our Sun, making it among the largest supergiant stars known to scientists. (You'll learn more about binary, dwarf, and supergiant stars in coming chapters.)

After Bessel's initial measurements, the number of stellar parallax studies gradually increased. Before 1900, astronomers had estimated the distances to about 60 stars using the parallax approach. The relative scarcity of good telescopes made it difficult for many researchers to accurately measure the small changes in star positions. In 1903, the American astronomer Frank Schlesinger (1871–1943) improved the accuracy of stellar parallax calculation by comparing photographs of stars taken at different times. A few years later, Schlesinger used long-focus refracting telescopes to make observations, and, in 1935, he published the parallaxes of over 7,000 stars.

Figure 1.6   Friedrich Wilhelm Bessel. German postage stamp issued on June 19, 1984, on his 200th birthday.

First, Schlesinger would photograph a relatively nearby star, carefully noting its exact position against the background of fainter, more distant stars. Half a year later, when the Earth had traveled 186 million miles to the opposite end of its orbit, he again photographed the same star. The shift of the star against the other background stars could be used to compute the star's distance. For increased accuracy, Schlesinger analyzed the photograph using a microscope and micrometer. By the 1920s, Schlesinger and other astronomers at large observatories knew the distances of about 2,200 stars using this approach.

* * *

Today we use several independent methods to measure the vast distances between stars in our Universe. Each method has limitations; for example, stellar parallax cannot be used to measure the distance to faraway stars. We're limited by the diameter of the Earth's orbit. Creatures on planets with very large orbits might be able to use parallax calculations to estimate locations of more distant stars. We can also use the *luminosity method* for measuring the brightness and periods of variable stars (such as Cepheids) and exploding stars (novae and supernovae), which you'll learn about in coming chapters. If these stars' absolute luminosity (energy output) is known, then their apparent brightness is related to their distances when the method is applied to stars in nearby galaxies. Another method, the *red shift method*, uses a spectrograph to analyze light from receding galaxies. Making use of the Doppler effect, and assuming that the recession velocity is related to distance, as indicated by the Big Bang theory of the expansion of the Universe, we can estimate interstellar distances.[9] (The Doppler effect, which explains why the color of a rapidly moving light appears to change, is discussed in chapter 3.)

We said that 61 Cygni is 10.3 light-years away, fairly close to Earth by stellar standards. The night star nearest to the Earth is a companion to Alpha Centauri (sen-TAUR-eye) known as Proxima Centauri. It has a parallax of 0.76 seconds of arc and is 1.3 parsecs (4.3 light-years) parsecs away. To get a feel for the immense distances involved, consider that our Galaxy is over 30,000 parsecs (100,000 light-years) across.[10] If the Earth's orbit is the size of a Ping-Pong ball, Alpha Centuari is 5 kilometers (3 miles) away, and the Sun is only 0.1 millimeters in diameter. Using these analogies, it's easy to see that outer space is like a vast ocean with tiny, insectile sailing ships spread very sparsely.

Given the large distances between the stars, even if one uses optimistic predictions for the possibility of advanced alien civilizations, the chances of an extraterrestrial race making *physical contact* with us is small. Astronomer Gerrit Verschuur of the Fiske Planetarium in Colorado believes that if extraterrestrial civilizations are, like ours, short-lived in comparison to the Galaxy's age, then there are probably no more than 10 or 20 of them in our Galaxy existing at this moment, each a lonely 2,000 light-years apart from one another. "We are," says Vershuur, "effectively alone in the Galaxy." This means it is very unlikely that UFOs and aliens are visiting us. In fact, C. S. Lewis (1898–1963), the Anglican lay theologian and novelist, proposed that the great distances separating intelligent life in the Universe are a form of divine quarantine: "The distances prevent the spiritual infection of a fallen species from spreading." If there is a Galactic club of aliens, perhaps it would be closer to the center of our Galaxy where the stars are more tightly packed, and the mean distance between stars is only one light-year instead of nine light-years as in our region of the Galaxy.

Our fastest spaceships can travel about one six-thousandth the speed of light. Our fastest ships would require 25,000 years to reach Proxima Centauri, the closest star. Radio messages would take decades to reach our neighbors and thousands of years to cross the Galaxy.

All this talk of comic loneliness, while humans live in a galaxy of hundreds of billions of stars, brings back haunting memories of Austrian poet Karl Kraus (1874–1936) who wrote, "One's need for loneliness is not satisfied if one sits at a table alone. There must be empty chairs as well." I am also reminded of the haunting lines from "Velvet Green," by the eighteenth-century British writer Jethro Tull:

> We'll dream as lovers under the stars:
> Of civilizations raging afar.
> And the ragged dawn breaks on your battle scars
> As you walk home cold and alone upon Velvet Green.

Forty years as an astronomer have not quelled my enthusiasm for lying outside after dark, staring up at the stars. It isn't only the beauty of the night sky that thrills me. It's the sense I have that some of those points of light are the home stars of beings not so different from us, daily cares and all, who look across space with wonder, just as we do.

— Frank Drake, *Is Anyone Out There?*

CHAPTER 2

The Joy and Paschen of Starlight

Though my soul may set in darkness, it will rise in perfect light;
I have loved the stars too fondly to be fearful of the night.

— Sara Williams, "The Old Astronomer to His Pupil"[1]

"**P**retty colors," Miss Muxdröözol says looking at Bob's latest artworks. Bob made these by throwing marine iceworms onto canvases of green paint. The worms' plump, pink bodies have maneuvered on the canvases using their setae, or body bristles.

As Miss Muxdröözol wriggles her long fingers near the glistening canvases, the artforms cast a kaleidoscope of turquoise reflections against the museum walls. She touches Bob's arm as she looks at the green glitter. "Psychedelic," she whispers. "Do you invite all your friends to your quarters to see this?"

Bob looks at her with one of his faces while his other face gazes into the pool of green reflections. "Not all, Miss Muxdröözol." Bob has a fractal corpus callosum, the structure that connects his brain hemispheres. He has no trouble seeing two scenes at once.

There is a sudden scratching noise. "Sir, I am ready for our next lesson on stars." Mr. Plex comes ambling in, his various appendages move in oddball synchronicity, giving him the gait of a drunken tarantula.

Bob feels a shiver go up his arms as he looks into Mr. Plex's shiny face. Bob feels a chill, an ambiguity, a creeping despair. It must be the hallucinogenic pheromones Mr. Plex sometimes emits. The effect usually wears off in less than a minute.

Bob looks into Mr. Plex's amber eyes, the irises haloed by fractured rings of crimson. No one moves. The scolex's eyes are bright, his toothy smile relentless and practiced. For a moment, the iceworm painting seems to wriggle with life. But when Bob shakes his head, the paintings are still. However, Mr. Plex remains. Cruel. Nightmarish. Bob feels like he is caught in a tomb of endothermic octapeds, and all the lights shut out.

Bob nods. "Mr. Plex, today I want to tell you how we find out what stars are made of. It is amazing that we know about the elements in stars even though they are separated from us by trillions of miles. It's all because starlight itself carries an incredible amount of information. In fact, starlight holds a secret code."

Miss Muxdröözol takes a step closer. "Bob, when I visited Earth and looked up at the night sky, I saw lots of stars. Any idea how many?"

Bob smiles, happy for an opportunity to impress Miss Muxdröözol with his candor, breadth of knowledge, and quick, incisive wit. "On a clear night, not too close to any big cities, you can see about 2,000 stars without a telescope. Earth's solar system is part of the Milky Way Galaxy with about 200 billion stars. The Milky Way revolves around its center, which probably contains a black hole."

"Black hole?"

"Scientists believe that black holes exist in the centers of some galaxies. These *galactic black holes* are collapsed stars having millions or even billions of times the mass of our Sun crammed into space no larger than our solar system. The gravitational field around such objects is so great that nothing, not even light, can escape from their tenacious grip. We'll talk about black holes later."

Bob approaches a flexscreen and sketches an ellipse to represent the Milky Way Galaxy (figure 2.1). "I've drawn Mr. Plex where our Solar System is located on one of the spiral arms of the Milky Way. Like all stars, the Sun moves through

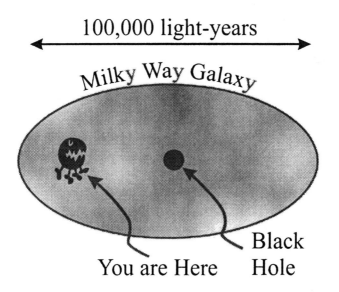

Figure 2.1   Milky Way Galaxy. Our Solar System is located at one end of the Milky Way Galaxy, where Mr. Plex is drawn. Perhaps ancient Church officials would have preferred the Earth to be located in a more central position, but today we should be happy we are not near the center in which a monstrous black hole resides with a mass more than 2 million times that of our Sun.

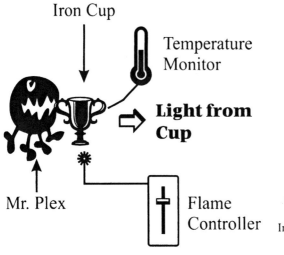

Figure 2.2   Glowing iron cup. The temperature of the cup determines which wavelength of radiated energy is brightest. Increase the flame temperature and the cup glows a different color.

space. It's fast—20 kilometers per second (45,000 miles/hour) with respect to nearby stars. Think of a sun as a Ferrari that drags along nine planets in its interstellar race. The Sun also moves in a nearly circular orbit—around the galactic center—with a speed of 220 kilometers per second!"

Bob tosses an iron cup to Mr. Plex who catches it in his mouth (figure 2.2). Next, Bob grabs a blowtorch and walks to Mr. Plex. "Don't worry. Your diamond body should shield you from the effects of the flame. Now watch. As I heat the cup, it first glows red, then yellow, and finally white. When it's white-hot, it also emits ultraviolet radiation that we can't see, but Mr. Plex can see it with his alien eyes."

Miss Muxdröözol pulls the blowtorch from Bob's hands. "What's the point of all this?"

"It turns out that the spectrum of colors an object radiates depends on several factors, which I'll teach you about. In some ways, a star is like the glowing iron cup in Mr. Plex's mouth."

Mr. Plex is starting to shake, perhaps from fear, so Bob grabs the iron cup with tongs and tosses it into a vat of water. "Similarly, stars radiate electromagnetic[2] energy. The hotter the star, the more energy it emits. The temperature of the star, like the iron cup, determines which wavelengths are brightest. I think I can explain this best with an illustration."

Bob speaks into the flexscreen, "Brunhilde, show solar spectrum," and the screen displays figure 2.3. "Our Sun emits a wide spectrum of radiation that peaks in the yellow range of the spectrum. On the flexscreen, you see a radiation curve for the Sun and the wavelengths at which the Sun radiates most intensely."

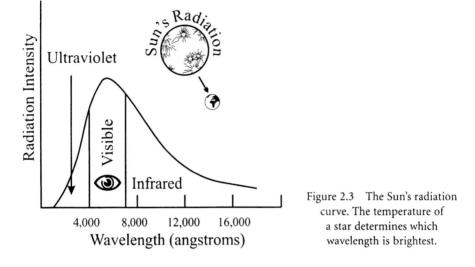

Figure 2.3   The Sun's radiation curve. The temperature of a star determines which wavelength is brightest.

Miss Muxdröözol stares at the screen. "It looks like the Sun radiates intensely in the visible range, but a lot of the energy is radiated at wavelengths we can't see" (figure 2.4).

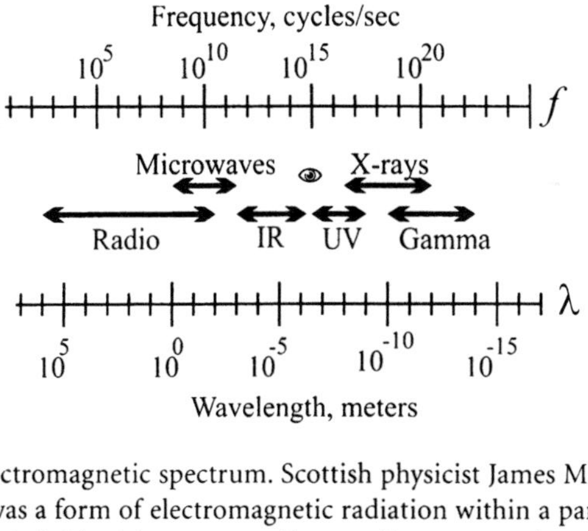

Figure 2.4   The electromagnetic spectrum. Scottish physicist James Maxwell (1831–1879) showed that light was a form of electromagnetic radiation within a particular wavelength range. All the waves labeled in this spectrum (from radio to gamma) are electromagnetic and travel at the speed of light. Many of the regions overlap; for example, one can produce radiation of wavelength 10^{-3} meters by both microwave and infrared techniques. The ☻ represents the visible range. [After Bob Halliday and Robert Resnick, *Physics* (New York: John Wiley & Sons, Inc., 1966), 993.]

"That's correct. Incidentally, by computing the area under the curve, you can estimate the total amount of energy the Sun radiates. Now I have a formula for you. We can calculate the wavelength lambda-max at which any star emits the greatest amount of radiation. Brunhilde, display radiation formula." On the flexscreen appears:

$$\lambda_{max} = \frac{0.3}{T}$$

"Here, lambda-max is the wavelength in centimeters and the temperature T is in degrees Kelvin (K)."[3]

Miss Muxdröözol runs her fingers lingeringly along the formula. "The hotter the temperature, the smaller the wavelength."

Bob nods. "Our Sun's surface is about 5,800 degrees Kelvin. 0.3 divided by 5,800 is 0.00005172 centimeters or 0.0000005172 meters or 5,172 angstroms (Å). Light that humans can see has wavelengths between 4,000 and 7,000 angstroms. An angstrom is very small, only 10^{-10} meters." Bob speaks into the flexscreen, "Brunhilde, show the electromagetic[4] piano," and the screen displays figure 2.5. "If we represent the electromagnetic spectrum as a 30-octave piano, where the radiation wavelength doubles with each octave, visible light only occupies part of one octave. If we wanted to represent the entire spectrum of radiation that has been detected by humans, we would need to add at least 20 octaves to the piano."

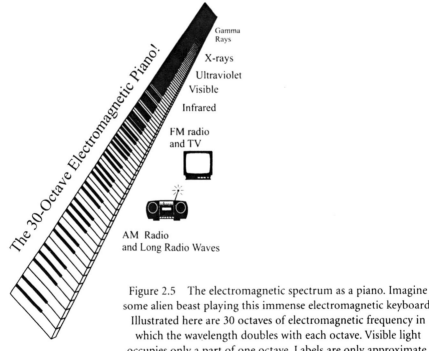

Figure 2.5   The electromagnetic spectrum as a piano. Imagine some alien beast playing this immense electromagnetic keyboard. Illustrated here are 30 octaves of electromagnetic frequency in which the wavelength doubles with each octave. Visible light occupies only a part of one octave. Labels are only approximate. See figure 2.4 for more precise locations. [After Denis Postle, *Fabric of the Universe* (New York: Crown, 1976), 59.]

One of Bob's faces stares at Mr. Plex, the other at Miss Muxdröözol. "I love the simple lambda-max formula because we can tell how hot a star is by its color (wavelength). The hottest stars look blue-white (short wavelength), and the coolest stars look red (long wavelength). For example, the hot star Vega, in the constellation Lyra the harp, is blue-white and has a surface temperature of 10,000 degrees Kelvin. On the other hand, the star Antares, in the constellation Scorpius, is 3,000 degrees Kelvin and red."

Mr. Plex nods. "Sir, the electromagnetic spectrum includes lots of wavelengths that we can't see. Do stars exist that radiate mostly in wavelengths that we can't see?"

"Yes, I think of them as ghost stars. For example, we can't see very cool stars with very long wavelengths."

Bob pauses as he walks over to a wooden desk in front of his large, mirror-backed bar. Bob stands in front of his android bartender, who comes to life when Bob sticks a finger in the android's solar plexus. "Hieronymus, I'll have some viper blood. Leave in the egg sacs." Hieronymus has been a good pal for several years, although Bob never quite got used to the funny Viking cap he seemed to enjoy wearing.

Hieronymus, the bartender

The android bows, "Excellent choice."

"Viper blood!" Miss Muxdröözol says.

Bob smiles. "You heard right."

"What is it?"

"A viper is a large, venomous snake—"

"I know what a viper is. What did you just order?"

"A drink with bourbon and a dash of viper blood aged together in charred oak."

"You're kidding?"

Bob simultaneously raises a single eyebrow on each of his faces. When the drink finally arrives, he notices a small, translucent sac floating on top of the liquid in the shot glass. "Good."

Miss Muxdröözol frowned. "Don't tell me what that is."

"It's a poison sac from a jumping pit viper, *Bothrops nummifera*." He drinks it down in one gulp. "Ow, that was hot." In a few seconds Bob breaks out in a sweat, a shiver runs up his back as if someone had just touched him, and his vision becomes spotty. Luckily, the effect lasts only a few seconds. He loves the feeling of living life on the edge, pushing himself to extremes, the belief that nothing can hurt him.

Miss Muxdröözol studies Bob's tearing eyes. "Do you really like that stuff?"

Bob takes a deep breath looking at her with one face than the other. "You don't have to love it. It's more important that it's unusual than that it tastes good." He pauses. "I'll give you a thousand dollars if you try it."

Miss Muxdröözol plays with her hair. "I don't know."

"Believe me, it's better than any man you've ever dated."

"Bob!" Miss Muxdröözol yells and opens her mouth wide.

Mr. Plex sits quietly through the entire exchange and remains expressionless. He asks the android for a raw calf's heart, harpoons it on a skewer, points it at Bob, and then pulls off a chewy piece of meat with his lips. "Sir, I hate to change the topic, but let's get back to the topic of stars."

"You are right. I want to finish our star lessons in a few days. Then I have a special treat for you regarding stars. So far we've seen that just by looking at stars we can often tell how far away they are and how hot they are. Ah, but we can tell so much more from starlight than that!" Bob tosses a ball to Mr. Plex. "The Danish physicist Niels Bohr, who lived from 1885–1962, represented the atom as a nucleus orbited by electrons. Think of that ball in your hand as an atomic nucleus and imagine electrons swimming around it at different distances from the ball's surface. Every element has different allowed orbits for the electrons. These orbits correspond to energy levels. Usually, an atom is found in its *ground state*, the state with the least energy. States with higher energies are called *excited states*. An atom in its ground energy can be excited to a higher state if energy is added by an amount that is equal to the difference between the two levels." Bob turns to the flexscreen, "Brunhilde, show energy levels." The flexscreeen displays figure 2.6.

"Sir, are the energy levels identical for atoms of the same type?"

"You bet. For example, allowed energies are the same for all atoms of oxygen with the same number of protons and neutrons. Now, take a look at the diagram. The center dot is the low energy ground state for hydrogen, which has one proton and one electron. When an electron falls from a higher energy state to a lower energy state, a photon of light is emitted. In the left of the figure, red light is emitted. At the right of the figure, blue light is emitted when the electron drops from the third energy level (or excited state) to the first excited state. In this case, because the electron undergoes a bigger drop in energy, it emits blue light that has a shorter wavelength and higher energy than red light. These might be observed as bright lines in an emission spectrum showing a range of the various possible colors of radiated light."

Miss Muxdröözol looks at the flexscreen. "Let me see if I got this right. Bright lines in atomic spectra occur when electrons jump from higher energy levels

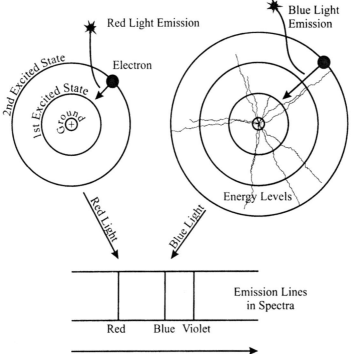

Figure 2.6 Emission lines in a spectrum (at bottom) produced by
electrons falling from high energy levels to lower energy levels.
Each chemical element has a unique emission line pattern
with colors of varying intensities.

down to lower energy levels. The color of the light depends on the energy difference between the energy levels. I've also heard of *dark* absorption lines in spectra that can occur when an atom *absorbs* light and the electron jumps to higher energy levels. The energy levels are identical for atoms of the same type."

Bob nods. "By looking at absorption or emission spectra we can tell what chemical elements produced the spectrum. In the 1800s, various scientists noticed that the spectrum of the Sun's electromagnetic radiation was not really a smooth curve from one color to the next but had numerous dark lines. This suggested that light was being absorbed at certain wavelengths. These dark lines are called Fraunhofer lines after the Bavarian physicist Joseph von Fraunhofer who lived from 1787 to 1826 and recorded such lines."

"Bob," says Miss Muxdröözol, "hold on a minute. I understand how the Sun can produce a radiation spectrum, like you showed us before, but what's this about dark lines? How can the Sun absorb its own light?"

"Think of stars as fiery gas balls containing lots of different atoms emitting light of all colors. Light from a star's surface, the *photosphere*, has a continuous spectrum of colors, but as the light travels through the star's outer atmosphere, some of the colors (light at different wavelengths) are absorbed. This is what produces the dark absorption lines. It reminds me a little of turning down the color intensities on a TV so that the picture becomes black and white. Except that with the sun, it's as if there are many different controls turning down many different narrow ranges of color. On a 'TV' of this sort, you'd still see a color picture, but certain shades of color might be missing. Imagine watching *Baywatch*, if all the red were suddenly absorbed, the bathing suits would turn black. If some odd color were absorbed, it might be hard to notice a change in the picture. In the stars, the missing colors, or dark absorption lines, tell us exactly what chemical elements are in the stars' outer atmosphere" (figure 2.7).

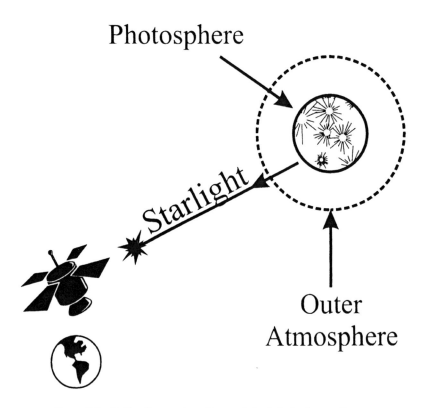

Figure 2.7   Stellar fingerprints. Some of the light in the continuous spectrum of energy from the star's photosphere is absorbed when the light passes through the star's outer atmosphere. The particular absorption pattern depends partly on elements in the star's atmosphere.

"Are there lots of elements in the Sun?"

"We've catalogued thousands of missing wavelengths in the Sun's spectrum. By comparing dark lines with spectral lines produced by chemical elements on Earth, astronomers have found over 70 elements in the Sun."

Bob walks over to a desk and withdraws three tiny robots. "Let me introduce to you my friends, Mr. Lyman, Mr. Balmer, and Mr. Paschen." Bob places them on the table.

Mr. Lyman   Mr. Balmer   Mr. Paschen

The three robots begin jumping up and down. Mr. Lyman seems the most energetic, jumping up to greater heights than the others before descending:

"Lyman Series"

"These jumping robots are metaphors for electrons in an atom that are hopping between allowed energy levels. Now, if I give one of these guys a poke with a finger, you'll see that they jump higher for a moment. To get a better feel for how all this works, let's look more carefully at a model of the electron's energy levels in a hydrogen atom. Brunhilde, display hydrogen atom energy levels." Figure 2.8 appears on the screen. "Here, we see a few more orbits or energy levels. You can think of the hydrogen's single electron orbiting its single proton, which is shown as a black dot at the center. Actually, the electron has an infinite number of orbits, of which the drawing just shows six. Upward jumps of the electrons produce absorption spectra. When electrons fall toward the ground state, numbered 1, they produce emission spectra. I haven't drawn the orbits' radii to correspond to energy differences, but the biggest energy difference is between the ground state, 1, and the second energy level, 2. The difference in energy between the higher levels is small."

"Sir, can I add a nearly infinite amount of energy to jump the electron to a nearly infinite energy level?"

"No. If the electron absorbs too much energy, it finally breaks free entirely of the proton, and the hydrogen is *ionized* leaving behind the proton, which is a positively charged particle."

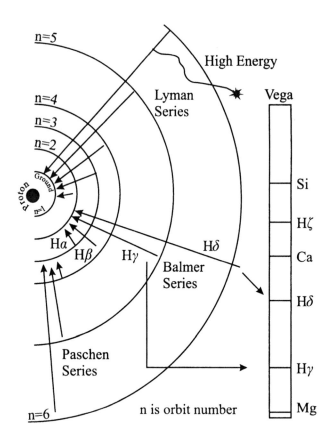

Figure 2.8   Atoms and stars. Energy levels for a hydrogen atom are depicted at left. The spectrum of the star Vega is shown at right. [After James Kaler, *Stars* (New York: Scientific American Library, 1992).]

Bob looks at Miss Muxdröözol. "Jumps that land on or come from the second orbit produce what is called the *Balmer series*, which correspond to spectral lines observable in optical spectra. The *Lyman series* correspond to more energetic changes and produces spectral lines in the ultraviolet. *Paschen* and higher order series produce low-energy infrared and even radio signals." He pronounces the last series PA-SHUN.

The little Lyman robot jumps off the desk. Bob picks him up and tosses him to the android bartender. "Hieronymus, see if you can figure out a way to turn Lyman off." Hieronymus struggles with the robot and finally gives up, crushing the robot beneath a bottle of rum.

"Scientists near the end of the 1800s studied emission of light from gases whose atoms are excited by electrical discharges. The resulting spectra are a set of lines that characterize the element. For example, scientists knew about the wavelengths corresponding to the lines in the spectrum of the simplest element, hydrogen. To be more precise scientists observed a series of lines at the following wavelengths in the visible range: 656.279, 486.133, 434.047, 410.174, 397.007, 388.905, 383.539, and 379.790 nanometers, or nm for short. (One meter is 10^9 nm or 10^{10} angstroms.) The series of lines continues to shorter wavelengths with the difference between successive wavelengths continually decreasing. For example, here are the visible Balmer lines in the spectrum of hydrogen. Brunhilde, display Balmer lines." Figure 2.9 is displayed. "In 1884, a Swiss school teacher, Johann Balmer found that the wavelength of a particular set of lines in the hydrogen spectrum can be predicted using a simple formula." Brunhilde, of her own volition, displays:

$$\lambda = 364.6 \frac{j^2}{j^2 - 4} \text{nm},$$

"Here, j is an integer that can have the values 3, 4, 5, . . . Lambda is the wavelength of each spectral line in units of nanometers (nm). Each value of lambda produced by the equation corresponds to a wavelength scientists observe."

"Bob, isn't it awesome that such a little formula captures a piece of reality?"

"You bet. And it was formulated by a secondary-school teacher in Basel, Switzerland, but Balmer never knew *why* the formula worked. We'll learn about the *why* in a moment. In 1913 Niels Bohr found that the Balmer formula supported his theory of discrete energy states within the hydrogen atom."

Miss Muxdröözol looks at Bob. "Ah, so that's where The Quadrapeds got the name of their song, 'Where have you gone, Johann Balmer?'" The Quadrepeds are a popular punk band comprised of software agents. The software breeds

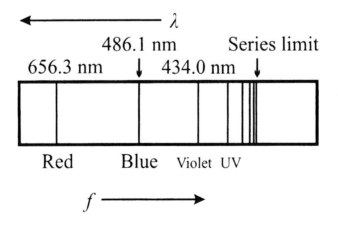

Figure 2.9   Balmer series for hydrogen. The wavelengths in nanometers for these lines may be calculated from $\lambda=364.6[j^2/(j^2-4)]$ for $j=3,4,5,\ldots$

within a supercomputer residing a mile beneath Antarctica. Hieronymus begins to sing some of the lyrics in his metallic voice: "Where have you gone, Johann Balmer? A nation turns its lonely eyes to you. Oooh, ooh, ooh."

Bob points a finger at Hieronymus and shouts, *"Wie alt sind Sie?"* Hieronymus stops signing.

"Eventually, this series of lines became known as the Balmer series. Balmer wondered whether his little formula might be extended to study the spectra of other elements. He knew similar patterns exist in the line spectra of many elements. He also wondered about spectral lines that the human eye can't see. A few years later, in 1906, additional series of lines were in fact discovered for hydrogen in the ultraviolet region of the spectrum. These were called the Lyman series after their discoverer, Theodore Lyman. Other famous series are the Paschen series, named after German scientist Friedrich Paschen, the Brackett series, named after U.S. scientist F. S. Brackett, and the wonderful Pfund series, named after U.S. scientist August Herman Pfund. The Paschen, Brackett, and Pfund series lie in the infrared region."

"Brunhilde, show transitions." Brunhilde displays figure 2.10. "The diagram indicates various series of transitions using arrows between the energy levels. As I suggested, the frequency of light emitted when an electron drops from one level to the other is related to the energy difference between levels. When Bohr wrote his paper in 1913, he only knew about the Balmer series and the Paschen series. In 1914, Lyman published a paper on his series, and in 1922 and 1924 Brackett and Pfund, respectively, published their series."

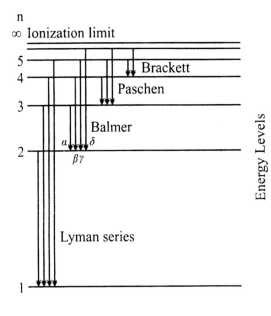

Figure 2.10   Energy-level diagram for hydrogen showing several transitions in the Lyman, Balmer, Paschen, and Brackett series. The Pfund series, not shown, corresponds to transitions to the fifth energy level.

Bob points to the various transitions in figure 2.10. "There is a more general formula for the hydrogen atom, proposed by W. Ritz that includes all four of these series." Brunhilde, of her own volition, shows the following formula:

$$\lambda = 91.14k^2 \frac{j^2}{j^2 - k^2} \text{ nm},$$

"In this equation, k is the integer 1 for the Lyman series, 2 for the Balmer series, 3 for the Paschen series, 4 for the Brackett series, and 5 for the Pfund series.[5] The integer j ranges upward starting from $k+1$. For example, if we substitute $k=2$ for the Balmer series, and $j=3$, the formula gives 656.21 nm, the wavelength of the line designated Hα, the first member of the series. The line appears in the red region of the spectrum. Atoms of other elements that have lost all their electrons except one also emit radiation that can be analyzed by formulas similar to Balmer's. In fact, this equation can be rewritten more generally so that it is applicable for other elements. Brunhilde, show general formula."

$$\frac{1}{\lambda} = R\left[\frac{1}{(j+a)^2} - \frac{1}{(l+b)^2}\right], \, l > j$$

"In this form, the equation is more commonly known as the Rydberg-Ritz formula. The constant R is the Rydberg constant, and for hydrogen it is one of the most accurately measured fundamental constants of the Universe (R_H = 10,967,758.1 ± 0.7 m⁻¹).[6] For very massive elements, R approaches[7] the value R_∞ = 10,973,731.568549 ± 0.000083 m⁻¹. R is the same for all series of the same element and varies only slightly in a regular way from element to element. The integers a and b are constants that are different for different series. The integers l and j are such that $l = j+1, j+2, j+3. \ldots$ Brunhilde, display table."

Table 2.1
Spectral Series

j	Name	Wavelength range
1	Lyman	ultraviolet
2	Balmer	near UV and visible
3	Paschen	infrared
4	Brackett	infrared
5	Pfund	infrared

Mr. Plex yawns. "Excuse me, Sir."

Bob walks over to the refrigerator. Inside is a bowl full of strawberries, one of the new genetically engineered varieties that turn translucent when cold. Each one also contains 50 milligrams of caffeine and other methylxanthines. Bob tosses a few to Miss Muxdröözol and Mr. Plex, who gobble them down with great gusto. Bob watches as Mr. Plex's irises contract and dilate at one-second intervals.

"Very good, Sir."

Miss Muxdröözol seems to be unaffected. Perhaps her body rapidly metabolizes the compounds to methyluric acids, or what is more probable is that she has no cyclic AMP phosphodiesterases with which the stimulants normally react.

Bob nods. "Not only does the Rydberg-Ritz equation give very accurate results for the hydrogen atom, but it represents accurately the line spectrum of many other atoms. Because the spectrum of an atom must be related in some way to its structure, the structure of atoms must be related to a series of integers."

Bob pauses and takes a deep breath. "Oh God, that idea gives me shivers. Why do these small integers rule the Universe? Can you imagine what problems we'd have if God had designed the Universe a bit more complexly? If there weren't such simple integer relations in atomic spectra, we might never be able to develop models for atoms."

"But Sir, these equations are not really models for atoms, are they?"

"Yes, they are in the sense that atoms can be modeled with electrons orbiting in different energy levels. Until 1913, scientists had not been able to devise any structure or model that accounted for this intriguing experimental correspondence of spectral lines with the Rydberg-Ritz formula. Scientists tried hard to create a model of the atom that would account for these formulas. Finally, in 1913 Neils Bohr proposed a model of the hydrogen atom that had wonderful success. His model had electrons moving in stationary orbits without radiating energy as they orbited. He said that the electron in a hydrogen atom occupies one of an infinite number of discrete orbits, each orbit being progressively farther from the nucleus and labeled with an integer. According to the Bohr model, the electrons only radiate (or absorb energy) when they jump from one stationary orbit to another. After a number of calculations,[8] Bohr was able to use his orbital theory for atoms to finally derive the Rydberg-Ritz formula and actually calculate the Rydberg constant based on various constants of physics."

Brunhilde displays:

$$R_H = \frac{m_e m_p}{m_e + m_p} \frac{e^4}{8c\varepsilon_0^2 h^3}$$

$$R_\infty = \frac{m_e e^4}{8c\varepsilon_0^2 h^3}$$

"Here, m_e is the electron mass, m_p the proton mass, e the electron charge, c the speed of light, ε_0 the permittivity of free space, and h is Planck's constant.[9] For now I don't want you to worry about what all these strange physical constants are. The point is that Bohr was able to come up with a simple atomic model that jibed with both R_H and R_∞. Today we know that his model of the hydrogen atom is simplistic, but it is nevertheless remarkably accurate in predicting hydrogen's energy levels. Take a look." Bob points to figure 2.10. "You can make a similar diagram yourself by positioning the lines (representing energy levels) on a vertical scale using $energy = -1/n^2$, where $n = 1,2,3,4 \ldots (n = 1$ is the ground state, and higher values of n are higher energy states). This is a bit simplified; the "1" in the numerator depends on the same kinds of physics parameters as did the Rydberg constant, but you can see how the spacing of energy levels becomes closer as n increases. The frequency of light emitted when an electron drops from one level to the other is related to the energy difference between levels."

Bob gives another few strawberries to Mr. Plex and Miss Muxdröözol. "Let's turn our attention back to our previous diagram of the electron orbits of hydrogen. I want to make sure you understand all of it." Brunhilde redisplays figure 2.8. "Let's review. There is a single proton in the center. One is the ground state. You can see six orbits total out of the infinite number of possible orbits. The optical absorption lines of Balmer arise from the second orbit. Each line has a name. For example, if energy is absorbed by an electron jumping from orbit 2 to orbit 3, the resulting hydrogen line is called Hα. A jump from orbit 2 to 4 is called Hβ. If we hope to see Balmer lines, there has to be a sufficient number of electrons in the second orbit, but that's no problem because nature seeks low energy states. This means we'll find lots of electrons here if the temperature is sufficiently high to move electrons to this level. In the cool atmosphere of an M star (like Barnard's star which is 3,000 degrees K), we don't usually see hydrogen lines because so few electrons are in the second orbit. We first begin to see hydrogen lines in class K stars at about 4,000 degrees K. Heavier atoms like calcium behave in similar ways as hydrogen, but the heavier atoms have more electrons and more complicated spectra."

Bob pauses and looks at Miss Muxdröözol, "Let me list some facts. Scientists have observed around 74 different elements in stars. We're sure other natural elements exist in stars, but their quantities may be too low to make their lines visible. The majority of stars have similar compositions, with hydrogen making up about 90 percent of the number of atoms, followed by helium at 9.9 percent. All the other chemical elements are in the remaining 0.1 percent. Generally, the more complex the atom, the less you'll find of it in stars. Incidentally, the stellar spectra help scientists measure the radial velocities of stars because the wavelengths of the lines are shifted slightly by the Doppler effect. We'll talk about that later."

Miss Muxdröözol is staring at the complicated looking Rydberg-Ritz equations. "Bob, helium is the second most common element in stars. Is it also the second most common element in the Universe?"

"Yes, it's second after hydrogen and it is the only element to be discovered on the Sun before it was discovered on Earth. It's light and doesn't combine with other elements to form compounds."

Bob hands Miss Muxdröözol a clear bag of marbles. "I like to think of the entire Universe as a great big bag of marbles so I can visualize how hydrogen makes up most of the Universe. This bag contains 10,000 marbles of different colors."

"It looks mostly blue."

"Yes, the blue marbles represent hydrogen. The marbles are in proportion to the kinds of atoms in the Universe. Of the 10,000 marbles, there are 9,017 marbles representing hydrogen, 976 marbles representing helium, 6 marbles representing oxygen, and 1 representing carbon. No other element would be represented by marbles in your cosmic bag."

"Helium was discovered first on the Sun?"

"Yes, the French astronomer Pierre Janssen detected a yellow line in the solar spectrum of light coming from the edge of the Sun. It was during an eclipse in 1868."

Mr. Plex stares at figure 2.8. "Sir, what's the diagram at right, marked 'Vega,' mean."

Bob glances at the flexscreen that displays figure 2.8. "We see a spectrum of the star Vega in which most of the electrons are in the second orbit, and some of the Balmer lines are quite evident. As I mentioned, Vega is one of the brightest stars in the night sky and the brightest star in the constellation Lyra. It's also the source of alien signals in the 1998 movie *Contact*. But the movie was unrealistic because it's unlikely aliens are actually there—Vega is too young (350 million years old) for intelligence to have had time to evolve."

Miss Muxdröözol comes a step closer to Bob. The air seems to be full of fragrance and murmurings until something slips by Bob's peripheral vision. "What?" he says gazing at the iceworm paintings. Is that something moving within the canvas, a snakelike thing that writhes back and forth and then is still? He touches the canvas, feeling along the base of the frame. His hand touches something that moves and scuttles off into the murky paint as he suddenly draws his hand away.

Bob's heart skips a beat. "Something's in there!"

Hieronymus and Mr. Plex gasp. Miss Muxdröözol opens her eyes wide and looks at the paintings. "What are you talking about?"

One of Bob's faces frowns, and then after a few seconds the other grins, although the smile doesn't reach his eyes. "I thought I saw something." But now he thinks the snakelike movement within the depths of the canvas must be just a trick of the light.

"This?" Miss Muxdröözol says as she moves her wrist that reflects a tangle of emerald reflections against the walls. Several gems are inlaid in her wrist. It is the latest fad, but call him old-fashioned, Bob prefers traditional jewelry— bracelets, rings, and chains.

"Yep."

"Bob, relax."

Bob looks at Mr. Plex to see what he is fiddling with at the bar. Mr. Plex is looking at a hologram of a girl Bob used to visit at art shows held in Martian caverns. The woman had a shaved area at the back of her head, in which a metal socket was surgically implanted. Mr. Plex must have picked up the hologram and looked at it at least a thousand times since Bob got it. Plex always put it back in the wrong place, when he was finished. You could tell he did it on purpose.

Bob looks away from Mr. Plex to a painting hanging near the bar. It is titled *Dance of the Esophagi*, depicting dozens of esophagi prancing in a field of wheat. Of course, it's as bizarre as paintings get. But it's also famous, by the artist formerly known as Flesh. Today the artist prefers to go by the symbol ⅃.

Bob takes a deep breath. Yessss . . . his mind plays funny tricks sometimes. *Stop it*, he thinks. Bob tries to focus on the enjoyment he feels explaining the stars to his friends.

Bob has a plan, a grand plan. Bob wants the Sun to become a beacon to all worlds, telling the rest of the Universe that intelligent life has evolved nearby. To provide evidence of their existence, humans should cloak the Sun in a cloud of material that absorbs some unusual wavelength of light. As a result, the entire Galaxy could view this magnificent beacon of intelligence. Astrophysicist Frank

Drake once suggested that an advanced civilization could place a technetium cloud around its star. This radioactive metal is observed on Earth only when it is produced artificially, and only weakly on the Sun, because it is short-lived and rapidly decays away. Drake estimated that aliens could mark a star using only a few hundred tons of light-absorbing substance spread around the star.

Bob has lofty dreams of making the Sun shine in strange ways. The odd stellar spectra will say to the Universe, "Humans are here!" The technetium seeding will require a few days to prepare, and Bob wants Miss Muxdröözol and Mr. Plex to fully appreciate the stars before he embarks on his audacious experiment.

Some Science Behind the Science Fiction

There are two ways of spreading light:
to be the candle or the mirror that reflects it.

— Edith Wharton, "Vesalius in Zante" in
Artemis to Actaeon

Bob's plan for engulfing the Sun in a technetium cloud is a daring one, isn't it? But his basic idea is sound. If we want to find evidence for the existence of extra-terrestrial civilizations, we must think of ways for detecting anomalies in space. We must look for information carried by electromagnetic radiation (radio, infrared, optical, ultraviolet, X-rays, and gamma rays), by cosmic ray particles (electrons and atomic nuclei), and by neutrinos. Some futurists have even suggested that information can be carried in gravitational waves, but this would be very difficult to detect given our current technology.[10] Other scientists have proposed that we search in the infrared region for beacon signals beamed at us.[11]

Around 1959, physicist Freeman Dyson proposed that we search for huge artificial biospheres created around stars by intelligent species (figure 2.11). This huge structure could be formed by a collection of alien-made habitats and planetoids that capture most of the radiant energy from the parent star.[12] Dyson suggests that the material from a Jupiter-sized planet could be used to create a large shell that surrounds the central star.[13] The shell would have a radius of one Astronomical Unit (AU), in which an AU is the mean distance between the Earth and the Sun—about 92,960,000 miles or 149,604,970 kilometers. The resulting "Dyson Sphere" would be a strong source of infrared radiation, peaking at ten micrometers. Humans have already attempted to find such constructs by searching for their infrared signals.[14]

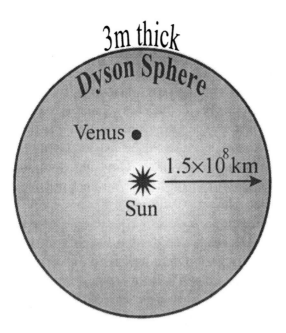

Figure 2.11   A Dyson sphere encompassing our Sun.

Returning to Bob's plan to use the element technetium to turn our star into an interstellar beacon, U.S. astrophysicist Frank Drake (b. 1930) and Russian astronomer Iosif Samuilovich Shklovsky (1916–1985) independently suggested that an advanced civilization could announce itself to the Universe by placing a short-lived isotope in the atmosphere of a star, an isotope that would not normally appear in the star's spectrum.[15]

Confused about all this talk of isotopes? The nuclei of isotopes of the same element have the same number of protons but have different numbers of neutrons. The isotopes of a given element have nearly identical chemical properties but different physical properties. A radioactive isotope has an unstable nucleus that decays. If we could scatter an unusual element around the Sun, the resultant "artificial smog" would produce a strong absorption line, and Drake proposed the use of technetium for this purpose. Technetium's most stable form decays radioactively within an average of 20 thousand years. Estimates suggest that we should dump 1.3×10^{11} grams (1.3×10^5 tons) of technetium into the stellar atmosphere (or into the stellar photosphere, the visible surface of the sun) to produce the beacon. So far, scientists have studied numerous sunlike stars and have not yet found technetium lines.[16]

In the late 1980s, French astrophysicist Hubert Reeves speculated on the origin of mysterious stars called "blue stragglers," first identified in 1952. These mysterious stars seem to have continued their hydrogen burning (fusion reactions) far beyond what would be expected of normal stars. In other words, they seem to be living extraordinarily long lives. When Reeves considered these stars, no one was quite sure what they were. Astronomers had long been mystified by observations of a few hot, bright, apparently young stars residing in well-established neighborhoods of older stars. Today we speculate that the stars result from mergers of one or more low-mass stars occurring in the dense stellar environment of globular clusters (a very dense community of stars). Their bluer color and greater brightness imply that they are more massive and much younger than normal globular cluster stars.

The mixing of the stellar material when multiple stars merge may increase the resultant star's lifetime. Stellar death normally occurs after the star leaves the "main sequence," a region on a luminosity versus temperature plot that hydrogen-burning stars inhabit. It's unlikely that our own Sun will someday turn into a blue straggler because in our area of the Milky Way stars are generally too far apart to be in danger of colliding. However, in the dense cores of globular star clusters, star collisions may be relatively common.

Although most astronomers feel there is a natural explanation for blue stragglers, astronomers like Reeves once speculated that some blue stragglers could have been created artificially.[17] For example, perhaps creatures on planets orbiting *artificial* blue stragglers could have found a way of keeping the stellar cores well-mixed with hydrogen that normally resides outside of the star's core, thus delaying stellar death after hydrogen core burning and the ultimately destructive red giant phase. (Normal stellar death is discussed in chapters 5 and 8.)

One way of mixing material far from the center of a star into its core is to explode hydrogen bombs inside the star, or to shoot lasers into the star, creating a "hot spot"[18] (figure 2.12). Various scientists have suggested that a similar mechanism might be used to increase a star's Main Sequence lifetime by a factor of ten or more. Incidentally, in the 1990s, astronomers discovered that the core of globular cluster 47 Tucanae is home to many blue stragglers, rejuvenated stars that glow with the blue light of young stars.[19] Figure 2.13 shows the crowded core of 47 Tucanae, located 15,000 light-years away in the southern constellation Tucana.

We've just discussed how blue stragglers might be created by stirring stars. Similarly, we could rejuvenate our own Sun so that it will support humans for a longer period of time by shooting lasers or hydrogen bombs into it. The precise reasoning is as follows.[20] As mentioned briefly in this book's introduction, the Sun obtains its energy by fusing hydrogen into helium in its super hot core. Right now, about 50 percent of the central hydrogen already has been con-

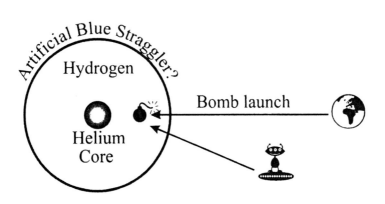

Figure 2.12   Artificial blue straggler star according to Hubert Reeves. The Earth launches hydrogen bombs or aims powerful laser beams at the Sun, creating a "hot point" and rejuvenating the unused hydrogen. This procedure could keep the star on the Main Sequence for a longer period of time than would be natural.

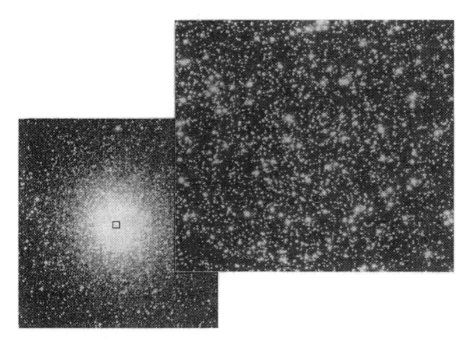

Figure 2.13   Blue stragglers in globular cluster 47 Tucanae. The ground telescope image at left shows the bright heart of the globular cluster. The image at right was taken by the Hubble Space Telescope, which is able to visually separate the dense clump of stars into many individual stars. If you look closely, you'll see several circles around stars that highlight several of the cluster's blue stragglers. A color image is available at http://oposite.stsci.edu/pubinfo/PR/97/35/a.html. [Figure courtesy of Rex Saffer (Villanova University), Dave Zurek (American Museum of Natural History), and NASA.]

verted to helium. In 5 billion years, there will be no hydrogen left in the hot core. At this point the Sun, devoid of fuel, will begin to choke and die. However, there is hope. Consider the huge amounts of unburned hydrogen between the core and the surface of the Sun. It seems to be going to waste. If it were possible to stir the hydrogen back into the core we could prolong the Sun's life from 10 billion to 100 billion years. And, as we've been discussing for the blue stragglers, one way to stir the Sun is to use hydrogen bombs or lasers. The challenge would be to have the bombs detonate *close* to the core without first vaporizing. I invite readers to invent imaginative solutions to this ultimate challenge.

* * *

In this chapter, we've discussed radiation from stars. To a first approximation, stars emit energy as a "blackbody," or theoretically as a perfect radiator. The hotter the star, the more energy it emits. As we learned, the temperature of a star determines what form of electromagnetic radiation is most intense. The German physicist Wilhelm Carl Werner Otto Fritz Franz Wien (1864–1928) won a Nobel Prize for work relating to his law that the maximum wavelength of energy emission for a blackbody is inversely proportional to the absolute temperature of the body, or, more precisely, λ_{max} (in centimeters) $= 0.3/T$. This formula makes clear the origins of a star's color. The color relates to different surface temperatures through the Wien blackbody law. Red stars are cool, near 3,000° K, yellow ones are closer to 6,000° K, white stars are about 10,000° K. Blue stars are really hot, around 20,000° K. As already mentioned, this means we can observe a star and estimate its temperature.

We also can predict the total energy a blackbody emits. According to the Stefan-Boltzmann radiation law formulated in 1879 by Austrian physicist Josef Stefan and derived in 1889 by Austrian physicist Ludwig Boltzmann, the amount of energy emitted is proportional to the fourth power of the temperature of the object. A star that is the same size and four times as hot as our Sun radiates 4^4 or 256 times more energy than the Sun. A spherical blackbody (like a star) will produce a luminosity, L, that depends on the star's surface area times the fourth power of its temperature. We'll discuss this further in chapter 5. We'll also discuss in more detail how the chemical composition of the stellar atmosphere can affect the appearance of certain giant stars.

The electromagnetic piano in figure 2.5 emphasized that humans can sense only a tiny portion of the electromagnetic spectrum. However, even though we are limited, it's possible that aliens exist on other worlds that have senses beyond our own.[21] Even on Earth, we find examples of creatures with increased sensitivities. For example, bats rely almost entirely on sonar; they hear the re-flection of their high-pitched sounds from their surroundings. *Gymnarchus* fish use weak electrical discharges to sense the locations of other fish. Members of the Mormyridae and Gymnarchidae fish families have electrical organs that allow the fish to distinguish prey, predators, members of their species, and ob-stacles in the water. Some animals appear to be sensitive to magnetism and radio waves. Rattlesnakes have infrared detectors that give them "heat pictures" of their surroundings. Given this diversity and range of senses on Earth, we can never precisely predict the nature of alien senses. As I noted in my book, *The Science of Aliens*, when I think about the possibility of traveling to alien worlds, I always remind myself that the most exotic journey would not be to see a thousand different worlds, but to see a single world through the eyes of a thou-sand different aliens. Not only do I mean this in the symbolic sense of viewing the world from various alien perspectives, but I also literally mean seeing through eyes sensitive to strange, non-visible parts of the electromagnetic spectrum, seeing in all directions simultaneously, or seeing events that are so quick that they are a mere blur to the human eye.

Aliens on worlds illuminated by a sun would have vision because of its sur-vival value. On Earth, eyes of various kinds had evolved numerous times in different animal groups. Even certain one-celled organisms have eyelike struc-tures called eyespots. In general, wherever a sense is vital to an animal, there is an accompanying, noticeable accouterment on the body surface. For example, if hearing is vital for an alien, I'd expect some form of ear that may be able to swivel in the direction of sound. If the sense of smell is important, we would expect a nose, long proboscis, or forward-projecting snout. On the other hand, if these senses are diminished or unimportant, and an alien relies on a sense like the electrical sense of fish, we would expect the external accouterments to be smaller with less features—eyes, ears, and noses would be smaller.

Life forms respond to the available input signals around them. This means that many aliens would have organs for sight because the Universe is often bathed in light. Aliens would also have some sense of touch so that they could respond to the physical world by moving around objects, avoiding dangerously sharp shapes, and so forth.

The tactile sense is probably the most basic of all senses, and there are very few life forms on Earth, however simple, that do not respond in some way to being touched. *Intelligent* aliens would have additional senses; it seems unlikely that a complex and intelligent creature could evolve with one sense alone. These additional senses are needed to confirm and refine an intelligent creature's per-ception of the immediate environment. This means that other senses are likely to accompany the sense of touch.

* * *

In this chapter, we've also discussed how integers in the Balmer and Rydberg-Ritz formulas do very well in explaining certain subatomic phenomena. It seems that there is something special about integers and the way they characterize the fabric of the Universe. Leopold Kronecker (1823–1891), a German algebraist and number theorist, once said, "The primary source of all mathematics are the integers." Since the time of Pythagoras, the role of integer ratios in musical scales has been widely appreciated. More importantly, integers have been crucial in the evolution of humanity's scientific understanding. For example, in the nineteenth century, British chemist John Dalton discovered that chemical compounds are composed of fixed proportions of elements corresponding to the ratio of small integers. This was very strong evidence for the existence of atoms. In the early 1900s, Balmer, Bohr, and others discovered that integer relations between the wavelengths of spectral lines emitted by excited atoms gave clues to atomic structure. The near-integer ratios of atomic weights was evidence that the atomic nucleus is made up of an integer number of similar nucleons (protons and neutrons). The deviations from integer ratios led to the discovery of elemental isotopes (variants with nearly identical chemical behavior but with different radioactive properties). Small divergences in the atomic weight of pure isotopes from exact integers confirmed Einstein's famous equation $E = mc^2$ and also the possibility of atomic bombs. Integers are everywhere in atomic physics. Integer relations are fundamental strands in the mathematical weave—or, as German mathematician Carl Friedrich Gauss said, "Mathematics is the queen of sciences—and number theory [the study of integers] is the queen of mathematics." In 1623, Galileo Galilei foretold the intertwining of mathematics and nature with his credo: "Nature's great book is written in mathematical symbols."[22]

Man is . . . related inextricably to all reality, known and unknowable . . . plankton, a shimmering phosphorescence on the sea and the spinning planets and an expanding universe, all bound together by the elastic string of time. It is advisable to look from the tidal pool to the stars and then back to the tide pool again.

— John Steinbeck,
The Log from the Sea of Cortez

CHAPTER 3

Spectral Classes, Temperatures, and Doppler Shifts

In the deathless boredom of the sidereal calm,
we cry with regret for a lost Sun.

— Jean de La Ville de Mirmont,
L'Horizon Chimerique

The next morning Bob slowly sits up in bed, pushes aside his blankets, and turns on *The Today Show*. The remote control is temporarily embedded in his right hand. Bob pushes a button, and one of the Picasso paintings on his wall suddenly transforms into a TV screen.

Bob listens with half an ear as Katie Couric interviews Madonna, the pop-singing sensation. Outside his window a few birds chirp in a cloudy morning air. Of course it is all fake. There are no birds, and the cloud scene is holographic. Katie Couric, the anchorwoman, died decades ago, but her mind was downloaded into silicon so she could continue to host her morning show. Madonna did not opt for electronic resurrection, but the gurus at Industrial Light and Magic created a robotic version indistinguishable from the original. The only difference, as far as Bob can tell, is Madonna's nose ring that functions as an antenna so that technicians can control her moves through radio signals.

People are never good at predicting where technology is heading. In the 1950s, there were those who thought we'd have nuclear-powered toasters in every home.[1] In the 1970s, few people predicted the Internet and all its ramifications. Humans think linearly, always trying to extrapolate from the present to the future, but it is the sharp turns that inevitably occur and lead to the most civilization-changing technology. Bob can barely imagine some of the uses scientists will find for the latest artificial people.

Bob likes his large bedroom in the museum ship. He looks out the windows at the stars. The more recognizable constellations are unmistakable. Orion's square shoulders and feet. The beautiful zigzagging Cassiopeia. The enigmatic Pleiades. They all remind him of life back on Earth where he studied the constellations long ago. He even sees Aldebaran, the red star in the constellation Taurus. He remembers the time he first saw Aldebaran as a boy growing up in Ajaccio, a town in Corsica, the birthplace of Napoleon. He moved to the United States when he was seven.

There is a knock at his door. "Sir, Miss Muxdröözol and I are ready for today's star lessons."

"One moment." Bob leaves his bed, jumps into his personal cleaning unit, and exactly $2\times\pi$ seconds later he is fully washed, dried, and clothed. He opens the door with a wave of his hand, and his two friends come in.

Miss Muxdröözol is wearing blue jeans—how retro!—and a beige T-shirt with a picture of Johann Balmer. She obviously likes Bob's lessons.

Bob studies her for a second and realizes he is fascinated by her every gesture, her way of speaking, and her smile and the way it crinkles her triple nostril nose. Bob wants to learn more about her culture and what her people are like. Obviously there is a vast chasm that separates Bob and Miss Muxdröözol. Could he ever truly have a deep friendship with someone who had never read, perhaps never even heard of, *War and Peace*, the Bible, or *Cryptonomicon*? She probably knew nothing about gourmet cooking, never traveled the Earth, never camped in a secluded New England pine forest. Great cultural cliffs separate creatures from different worlds.

"What the hell is that?" Miss Muxdröözol says, gesturing to one of the walls of Bob's bedroom. Along the wall hang daggers, sabers, rapiers, scimitars, wide-bladed knives called misericords, and other instruments of mayhem, the function of which most people could not quite discern. These are remnants of Bob's passion for Shaolin Kung Fu. These days he prefers the flowing, mystical motions of Tai Chi over the more obvious forms of martial art.

"Just a hobby," Bob says. "Come, sit by the table. Today we will talk about the different *spectral classes* of stars. Look out the window. Each of those stars has a different spectrum with different absorption lines. We discussed some of that yesterday. Hydrogen lines are more obvious in the stellar spectra for some stars than in the Sun's spectrum. Astronomers once thought that the stars with stronger hydrogen lines had more hydrogen than those with weaker hydrogen lines, but today we know that all the visible stars contain very similar elements in them, and the stars are made mostly of hydrogen and helium. Anglo-American astronomer Cecilia Payne-Gaposchkin (1900–1979) proved that differences in star *temperatures* cause most of the spectral differences in the dark line patterns. Before her, U.S. astronomer Annie J. Cannon (1863–1941) examined the spectra of thousands of stars and classified stars in seven spectral classes: O, B, A, F, G, K and M."

Miss Muxdröözol smiles as she strokes her silky, rainbow-colored hair. "Cecilia is one of my heroes. I have a photo of her in my bedroom. She was the first person, either male or female, to get a Ph.D. in astronomy from work done at the Harvard College Observatory."

Bob raises his eyebrow, thinks for a moment, and says, "Impressive that you should know such obscure facts. But let me tell you about the spectral classes. The classes represent temperature categories. The coolest stars are M stars. The hottest are O stars. Each spectral class has ten subclasses numbered zero to nine that indicate finer temperature gradations."

Hopefully Bob has impressed Miss Muxdröözol with his barrage of facts. He decides to continue. "The less massive the star, the longer it lives. The average star in the Milky Way is about half the Sun's mass and lives about 50 billion years." Bob speaks into his flexscreen, "Brunhilde, show spectral class," and the following table is displayed.

Table 3.1
Stellar Spectral Classes[2]

Spectral Class	Temperature (Kelvin)	Spectral Characteristics	Color	Mass (Sun=1)	Approximate life (years)
O	> 30,000 (Hot!)	Lines of ionized helium; few lines	Blue	40	1 million
B	10,000–30,000	Lines of neutral helium	Blue	10	20 million
A	7,500–10,000	Very strong hydrogen lines	Blue-white	5	400 million
F	6,000–7,500	Strong hydrogen & ionized calcium lines	White	2	2 billion
G	5,000–6,000	Strong ionized calcium and iron lines	Yellow	1	14 billion
K	3,500–5,000	Strong lines of neutral metals	Orange-red	0.5	50 billion
M	< 3,500 (Cold)	Bands of titanium oxide molecules	Red	0.2	140 billion

Mr. Plex studies the table, "Sir, as you suggested before, this means that you can often look at a star's spectrum and tell how hot the star is."

"Yes. And let me impress you with my knowledge of a few other classes. You don't need to know these minutiae for the rest of our lessons, but there are some less common classes, such as WC or WN class stars (Wolf-Rayet stars— white stars that are hot like the O class stars), C class stars (dim red stars probably too cool to support life), and S class stars, which are cool like the M-class stars and reddish-brown in color. N stars are also not part of the standard spectral sequence. They differ from the M stars not by temperature but by composition, having a carbon-to-oxygen ratio reversed from M stars. R stars are warmer carbon-rich versions and have the same temperatures of K and G stars. For now, don't let all these letters confuse you, just focus on the stars in the table."

Miss Muxdröözol nods. "The Wolf-Rayet stars sound interesting. What do we know about them?"

"They're named after their discoverers, the French astronomers Charles-Joseph-Étienne Wolf and Georges-Antoine-Pons Rayet. Humans have known about them since the late 1800s. The Wolf-Rayet stars are extremely hot and have spectra that suggest great turbulence within the star or a steady and large expulsion of stellar material. The Wolf-Rayet stars are usually several times the diameter of the Sun and thousands of times more luminous. These stars might be the central engines of many planetary nebulae, wispy expanding shells of gas surrounding a star's core."

"Planetary nebulae?"

"We'll talk more about them soon. Planetary nebulae are huge clouds of glowing gas that certain kinds of stars emit near death. The word "planetary" is used because their shapes reminded eighteenth-century astronomers, using primitive telescopes, of planetary disks. High-mass stars have the firepower necessary to heat their nebulas enough to glow during the brief time such gas clouds exist, about 10,000 years or so."

"So Wolf-Rayet stars are hot?"

"Very. 25–50,000+ degrees Kelvin. They're also massive stars—most range from 10 to 30 solar masses—with a high rate of mass loss. Strong, broad emission lines (with equivalent widths up to 1000Å!) arise from the winds of material being blown off the stars. *Wolf-Rayet galaxies* are galaxies that contain a large population of Wolf-Rayet stars."

Mr. Plex scratches his head. "So what do they look like?"

"Brunhilde, display Wolf-Rayet stars." Figures 3.1 through 3.3 are displayed. "This wind from the star becomes so thick that it totally obscures the star. This means that when we look at a Wolf-Rayet star, we're really just seeing this thick wind. The wind carries about the same mass of the Earth into space each year. The star is losing so much mass that it doesn't live as long as a normal star. Very

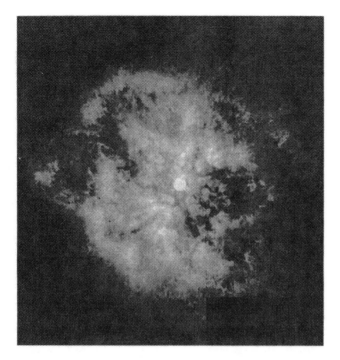

Figure 3.1   Resembling a fireworks explosion, this stunning NASA Hubble Space Telescope picture of the energetic star WR124 reveals the surrounding hot clumps of gas being ejected into space at speeds of over 100,000 miles per hour. The massive, hot central star is known as a Wolf-Rayet star. This is a very rare and short-lived class of superhot stars (in this case, 50,000 Kelvin). [Figure courtesy of Yves Grosdidier (Université de Montreal and Observatoire de Strasbourg), Anthony Moffat (Université de Montreal), Gilles Joncas (Université Laval), Agnès Acker (Observatoire de Strasbourg), and NASA, November, 1998 http://oposite.stsci.edu/pubinfo/pr/1998/38/ pr-photos.html]

massive stars may become Wolf-Rayet stars just before they explode as supernovae. Their mass loss seems to be so dramatic that they have lost their entire hydrogen envelopes."

"Are these stars rare?" Miss Muxdröözol asks.

"We only know about a few hundred of them, and they are mostly in the spiral arms of the Milky Way."

"Sir, maybe you and Miss Muxdröözol can take your honeymoon vacation around one of the Wolf-Rayet stars. That should really heat things up."

Bob looks at Mr. Plex with one face while his other studies Miss Muxdröözol. "Mr. Plex, what are you talking about! I hardly know Miss Muxdröözol."

"Sir, Hieronymus, your bartender, told me that you were getting married."

Bob walks over to Hieronymus and reactivates him by poking the solar plexus of the android. "Sir, how may I serve you?"

Figure 3.2 Detail of the star
WR124 and the surrounding
nebula M1-67, as seen in
figure 3.1.

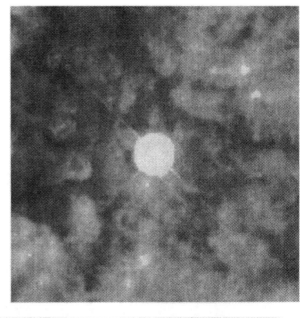

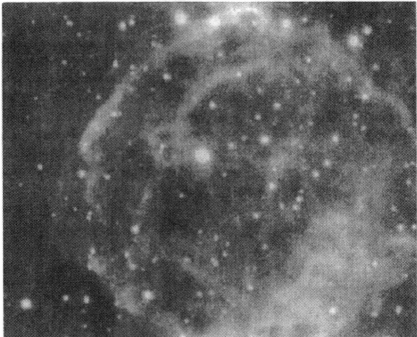

Figure 3.3 Wolf-Rayet star, the brightest star in the center of the image. This star, located in
the nebula known as NGC 2359, is surrounded by the large bubble of gas that is flowing from
the star's body. [Figure courtsey of Perry Berlind and Peter Challis, Harvard-Smithsonian
Center for Astrophysics (CfA), 1.2-meter telescope, Fred Whipple Observatory,
http://nmp.jpl.nasa.gov/st3/st3images/WolfRayetBig.html]

"Did you mention a honeymoon to Mr. Plex?"

"I was just kidding, Sir."

Miss Muxdröözol takes a deep breath. "I thought androids never kid."

"Almost never," Hieronymus says as Bob pokes his hand into the android's belly again to deactivate him. For several seconds Hieronymus's eyes dilate, and then they slowly close like a pair of dying flowers.

Bob shakes his head. "Pay no attention to Hieronymus."

Mr. Plex looks at the table of stellar spectral classes. "But Sir, you said that stars are mostly hydrogen and helium. Why such different spectra if stars have mostly the same elements?"

"It's the temperature that matters. Different atoms have different temperatures at which they produce spectral lines most efficiently. Also, in the superhot O stars, atoms are actually *ionized*."

"Ionized?" Mr. Plex says.

"*Mein Gott*, Mr. Plex. Haven't you learned anything in your chemistry class? A neutral atom has as many protons as electrons, but the electrons are only loosely held to the atom. If an atom is supplied with enough energy, electrons can be lost to produce positively charged ions. For example, carbon with one electron gone is symbolized as C^+. Now, as I was saying, in the hottest stars, only the most tightly bound atoms, such as singly ionized helium, survive, and the lines of ionized atoms are seen in the stellar spectrum. At cooler temperatures, below 3,500° K, entire molecules can survive. One example is the titanium oxide molecule."

Miss Muxdröözol sits back in Bob's massaging, high-back, executive leather chair. She smiles as the five built-in battery-powered massaging units begin to work on her clavicle areas. For a moment, Bob's mind wanders as he thinks of the chair's pneumatic gas lift, providing instant seat height adjustment.

Bob watches her for another second as she places one of her mottled appendages on the durable polypropylene armrests. "What class is the Sun?" she says in a low voice.

Bob pauses to catch his breath. "The sun is a class G star, with a temperature around 5,800° K. Neutral metal atoms like iron and nickel survive in our Sun. By the way, you can also see from the table displayed on the flexscreen that the stellar classes also indicate a color sequence from blue (the hottest O-type stars) to red (the coolest M-type stars)."

Miss Muxdröözol begins to evaginate her digestive tract through her oral cavity in an apparent expression of heightened interest. "It seems to me," she says, "that the absence of an absorption line in a star's spectrum doesn't necessarily mean that the star doesn't have the element."

Bob sits down on his waterbed. "You are perceptive as always. The main point of today's lesson is that the star's *temperature* influences the spectrum. The temperature determines which atoms are capable of producing absorption lines that we scientists on Earth can see."

Bob thinks he sees a movement in his room, something creeping along the walls, but when he blinks everything is normal. He gazes up at his lithograph of Picasso's *Les Demoiselles d'Avignon*, which gives the room a primitive, exotic sort of feeling.

Bob thinks he hears a soft sound coming from somewhere else. "Now what?" he whispers. "Can everyone be quiet for a moment?"

There is a quiet hissing sound, like the sound of a broom on wet cement. *Smells like an ocean at low tide*, he thinks. *Smells like crawling things.* His eyes nervously dart around the room searching for the source of the sound and smell.

Mr. Plex steps closer. "Everything okay, Sir?"

Bob nods. "I haven't been sleeping that well. Everything's okay. Let me finish up today's lessons by teaching you about stellar motions. Stars are moving through space with respect to the Sun, although these motions may not be obvious to us given our relatively short lifespans."

"How fast?"

"Many kilometers a second. A star's motion is called its *space velocity*." Bob pauses. "Miss Muxdröözol, may I borrow your lipstick?"

Miss Muxdröözol tosses him red lipstick in a stiletto-shaped sheath, and Bob begins to sketch on his waterbed in order to get his friends' attention. "Space velocity has two components because a star moves toward or away from us along our line of sight, and it also moves at an angle to us. This means that the space motion can be thought of as having two components, the *radial component* and the *proper motion component*. The proper motion vector is perpendicular to the line of sight, along which lies the radial component."

Bob points to the arrow in his diagram that represents radial velocity (figure 3.4). "Now, we can calculate the star's radial velocity simply by looking at its stellar spectrum. If the star is moving toward us, its light is *blueshifted*; that is, its spectral lines are shifted to shorter wavelengths. If the star is moving away from us, its light is *redshifted* to longer wavelengths. All you have to do, for elements on Earth, is compare a star's spectral lines with corresponding lines that represent electron transition wavelengths. For example, you can look at a reference spectrum of iron produced on Earth and compare that to the iron line in a star's spectrum." Bob writes out the phrase

Doppler shift

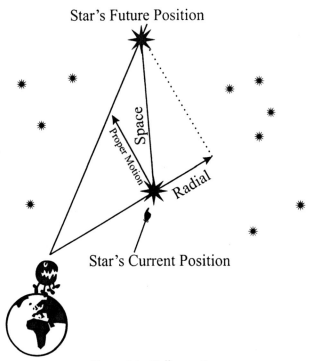

Figure 3.4   Stellar motion

on the waterbed, hoping Miss Muxdröözol takes note of his wit, penmanship, and erudition. "This shifting of wavelengths is called the *Doppler shift* after the nineteenth-century Austrian physicist Christian Doppler who found that the effect applied to many kinds of waves."

"Explain," says Miss Muxdröözol.

"Certainly. Let's go for a walk as we talk." Miss Muxdröözol takes her lipstick while Bob grabs a portable flexscreen. They all leave Bob's living quarters and walk down a little-known corridor to the Gallery of Odonta. All of the paintings in the gallery depict aliens with long bodies and two narrow pairs of intricately veined, membranous wings. Bob's favorite image is titled, "The Banded Demoiselle." It depicts an odd creature, colored electric blue, in conversation with a metallic green companion with broken wings.

Bob can study more than one painting at a time without moving his head. He simply shifts his attention from his one set of eyes to the other. Quite convenient.

"When a moving source emits waves," Bob says, "the wavelengths of the waves are changed for a stationary observer. Brunhilde, show Doppler shift." The flexscreen shows a stellar spectrum Doppler shift (figure 3.5). "In the figure,

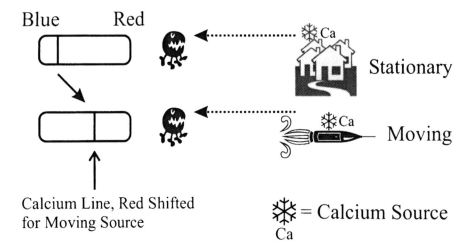

Figure 3.5   Doppler effect with calcium source, indicated by ❋. The calcium line
in the spectrum at top is redshifted in the bottom spectrum because the source is moving
away from Mr. Plex. For stars moving tens of thousands of miles per second away from us,
the redshift can be quite pronounced.

I'm showing a calcium source because stars usually show calcium lines in their
spectra. The bottom spectra indicate that the calcium lines occur at longer wavelengths than those from a stationary terrestrial light source containing calcium.
Using this kind of spectral evidence, we can tell that the galaxies are receding
from us, and the recession velocity is greater for more distant galaxies, which
suggests our Universe is expanding. For example, measurements of Doppler shifts
in calcium lines for nebulas in the constellation Corona Borealis[3] suggest that the
galaxy is moving away from us as fast as 13,400 miles per second."[4]

    Mr. Plex grabs the lipstick from Miss Muxdröözol's hand and sketches the
following formula on the museum wall:

$$\frac{\Delta\lambda}{\lambda} = \frac{v}{c}$$

"Sir, I have studied the Doppler effect. The change in wavelength $\Delta\lambda$ divided
by the wavelength from a stationary source (λ) equals the space velocity (relative velocity) v divided by the speed of light c."

    "Mr. Plex, please do not be so rude as to interrupt my discussions. But, I
must admit, that you are essentially correct. You can use the formula so long as
v is not too close to the speed of light. In other words, your formula shows that
the faster (v) a star is moving away from us, the greater the wavelength its emitted light gets shifted ($\Delta\lambda$). The main lesson I want you to remember today is
that a star's spectrum tells us about the star's radial velocity."

Mr. Plex begins to bob up and down. "*Mon Dieu.* Starlight tells us so much!"

"Mr. Plex! There is a lady present. Please contain your enthusiasm."

Miss Muxdröözol's piezoelectric eyelashes flutter like the wings of Rigelian moths, and Bob smiles. "Okay, so far we know how to determine radial velocity, but what about proper motion? Take a look at my previous figure (figure 3.4) before you answer."

A hush comes over Mr. Plex and Miss Muxdröözol.

"Proper motion is very gradual. We must measure it over several decades. In fact, the average proper motion for all visible stars is less than a tenth of a second of arc per year. At this sluggish rate, you won't notice any change in the appearance of constellations like the Big Dipper during our lifetimes, but 40 or 50 thousand years from now, the Big Dipper would look like a horse. The reason: In 50,000 years the angular change in an average star's position (as seen from Earth) is 5,000 seconds of arc or 1.4 degrees. This means each star will move about three times the Moon's angular diameter, which is half a degree" (figure 3.6).

Bob and his friends walk to a painting of the Earth with an *Orthetrum cancellatum* hovering beside it. A black tail protrudes from the alien's abdomen and appears to be inserted into Antarctica. A grin is on the creature's face. At least Bob thinks it's a grin.

"Sir, the Earth is upside down. Antarctica is on top."

"Mr. Plex, ever wonder why all modern maps of Earth show the northern continents like Europe on the top and the southern continents on bottom?"

Mr. Plex nods, his brachiocephalic veins pulsing in synchrony with his extended head motions.

"It's because the great map-making civilizations developed in Earth's Northern Hemisphere. Why would they want to put all the world they knew at the bottom? This bias is called *ethnocentric geography* or *hemicentrism*. If civiliza-

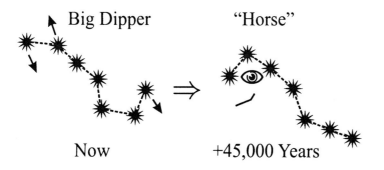

Figure 3.6   Dipper disintegration. In 50,000 years from now, the Big Dipper constellation will look like a horse. Proper motion of a few stars is indicated by arrows.

tion in the Southern Hemisphere had first grown in size and sophistication, and developed mapping skills, then today what we think of as south would be at the top of maps, and compass needles would 'point' south instead of north."

Mr. Plex and Miss Muxdröözol go closer to the painting of the "upside down" Earth, and then Miss Muxdröözol screams. Apparently, the smiling *Orthetrum cancellatum* grins even wider—evidence that the painting is one of those new forms of artwork with a proximity sensor. The closer a viewer goes towards the painting, the wider the grin becomes. As the creature smiles, Antarctica shrinks. It's as if the winged creature is draining the life out of a helpless captive.

Miss Muxdröözol takes a deep breath. "I can think of other things that would change if your dominant civilizations first appeared in what you call the Southern Hemisphere. The names of the constellations would be different.[5] For example, viewed from Australia, Orion the Hunter is standing on his head—hardly a heroic pose! The direction clockwise would be opposite from how you now define it. Your human clocks rotate as does the shadow in a sundial in the Northern Hemisphere. In the Southern Hemisphere, the sundial's shadow is counterclockwise."

"Very true, Miss Muxdröözol. Also note that the art of navigation would have developed more slowly in the Southern Hemisphere since there is no equivalent to Polaris, the North Star, in the southern skies.[6] Polaris appears to stand still, while other stars rotate around it, because it is almost exactly aligned along the celestial North Pole. This property is useful to navigators, because the star always lies in the direction of north."

Bob looks out a window at the stars. He finds himself doing that a lot lately. There is something about them—something ancient yet dynamic. He finds it calming to look at them, to know that there are things that endure not just for centuries but for millennia.

Now Bob speaks in a quiet voice to get his friends' attention. "Listen. This is what makes me shiver. Much more information comes from starlight. Even the *shape* of each spectral line gives information. The amount the lines broaden can give clues about the velocity of the emitting gas and the rotation of a star around its axis. The amount of splitting or broadening of lines can tell us about the strength of a magnetic field. We can't see this by looking at the star with our naked eyes. Mr. Plex, not even you can see it with your multiple photoreceptors. But scientists on Earth can use sensitive spectrometer equipment to study the spectral line shapes."

Miss Muxdröözol taps Bob on the arm. "This is so much to digest. Let's take a break."

Mr. Plex nods. "Sir, Miss Muxdröözol and I need to retire for a few minutes to recharge our esophageal implants and sphincters."

"Okay, I'll come get you in a half hour."

A few minutes later, Bob is sitting on his bed, paging through a glossy magazine called *Odobenus rosmarus*. It features pornography from several star systems, with an emphasis on nude heavy-bodied creatures sporting rounded heads and small eyes. Their muzzles are long and covered with stiff, quill-like whiskers. Bob didn't buy the magazine for the startling photos (the females are twice as large as the males, when fully hydrated)—he prefers the political articles in the magazine.

Suddenly, Bob hears a noise. "What the hell?" He feels the sharp jackhammer movements of his heart growing as fear closes a vise-like grip on his chest. Bob sits up in bed and looks directly at the source of the noise. *Am I dreaming? Hallucinating?*

Lately, bizarre dreams have been troubling his sleep, and sometimes his nighttime dreams ooze into the sunlight hours. Every now and then he has difficulty distinguishing dream from reality. *I'm not running from the bedroom*, he thinks. *That thing in the corner is just a fragment from a fading dream.*

But the sweeping noise continues. Bob cautiously reaches over to his bedside lamp and increases its intensity, expecting the apparition to disappear in the bright light. *It isn't possible!* The thing in the corner doesn't vanish. He nervously rubs his hands together. The light only makes it more real, more horrifying.

Sweat forms on Bob's foreheads. What is it? With a trembling hand, he draws the bed covers around him, and then he peeks.

He stares at a membranous curtain of geometric shapes. A single eye sits in the center. Thin veins pulse on the translucent curtain.

The membrane undulates slightly as it scrapes the carpeting of his room, and the eye occasionally blinks and swivels in his direction. It is a wild, exotic sapphire eye—utterly alien with long gray eyelashes.

At first Bob thinks the veined membrane around the eye is just a collection of shadows. But after closer inspection, he sees that the membrane is solid and pulsating.

The light from his lamp seems to be swallowed up by the membrane before him. He holds his breath. The large eye reminds him of an insect eye. This impression is enhanced by its many tiny facets. Every few seconds the eye jitters, and then the veins wander over the membrane like worms on an infected scalp. Some move in straight lines, others move in small spiral eddies.

At first, Bob is afraid to move a muscle. It reminds him of when he was a child and woke up in the middle of the night knowing that a monster was hiding under his bed or in the closet. Now, he feels like hiding under the covers like a child. But his adult mind soon takes control and repeats to him in no uncertain words:

Get out of the room.

Get out of the room, now!

Still Bob is afraid to move. His body does not respond.

Woosh. Woosh. From behind the membrane rises an undulating gelatinous mass resembling a small naked brain. A nerve cord of some thick fibrous material trails from the base of the brain to a connection on the membrane.

Get out! Bob's inner voice implores again.

"Oh my God, what is it?" Hieronymus begins to scream. *How had he become activated?*

Bob looks at Hieronymus with one face and the creature with the other. As he watches, some of Bob's fear gradually transforms itself into rage—rage that this thing would dare invade his bedroom, rage at his own fear and impotence.

A shot of adrenaline courses through his body as he emerges from the bed covers. With a quick intake of breath like someone about to plunge into icy water, Bob darts for the misericord hanging on his wall. With the knife in his hands, he runs toward the membrane, swings the knife with all his might, and cleaves the creature into two pieces.

But the creature still lives, and the two smaller pieces seems more energetic, the severed pieces of veins creeping and crawling like hornets in a disturbed hive. Both pieces of the creature come toward him.

"See you in hell!" Bob screams as he throws the misericord at the creature, slams the bedroom door shut, turns to his right, and runs toward the front of his living quarters. In the dim light the green shag carpet looks like the living tentacles of a grassy swamp thing.

Without looking back, he runs through the hallway of his quarters. The vibrations from the slammed door caused his Miró lithograph to fall from the wall and the fish in his nearby aquarium to scurry for cover behind some aquatic plants. His saltwater pycnogonid jams itself behind a rock. It is shielding itself from the commotion outside the confines of its glass-enclosed universe.

Bob trips over what looks like a giant spider and sprawls on to his carpet. *What next?* He gets up and runs for the door to his living quarters. As he puts his hand on the door knob he turns around to look at the brown thing on the floor and realizes that it is just the large pillow from his couch.

In the distance he hears Hieronymus screaming. There is an odd burning smell and ethereal sounds, something between a sighing harp and a xylophone.

Some Science Behind the Science Fiction

*One winter evening my Mother was wheeling me in my pram, and we saw
a brilliant meteorite blaze across the sky . . . She . . . taught me the right
name for it by making a little rhyme: 'As we were walking home that night,
we saw a shining meteorite.' It was my first encounter with astronomy.*

— Cecilia Payne-Gaposchkin,
    An Autobiography and Other Recollections

In this chapter, we discussed the movement of stars through space. As also indicated in chapter 2, stellar motions can be quite complex when considered in their totality. For example, our Sun actually orbits the center of the Galaxy every 250 million years, dragging Earth along for the ride. This means the Sun's speed is about 140 miles per second. The last time that the Sun passed at this exact place in the Galaxy, dinosaurs ruled the Earth. The Milky Way is part of a small cluster of galaxies called the Local Group. We fall toward the center of the group at about 25 miles per second. The entire galactic neighborhood is moving about 370 miles per second towards the constellation Hydra. This motion may result from the gravitational pull of vast superclusters or walls of Galaxies—the biggest known structures in the Universe. The Universe is a realm of perpetual motion where everything rotates around something else.[7]

We've also discussed how stars are categorized according to their spectral class. The origins of some of the classes date as far back as the late 1880s when Father Angelo Secchi (1818–1878), official astronomer to Pope Pius IX, conducted some of the early analysis of starlight using primitive spectroscopes.[8] These spectral classes were originally denoted O, B, A, F, G, K, M, which can be remembered by saying, "Oh, Be A Fine Girl/Guy, Kiss Me." In recent years, the classes have expanded slightly to O, B, A, F, G, K, M, R, N, and S ("Oh Be A Fine Girl, Kiss Me Right Now, Sweet!"). Both systems classify stars according to their temperature and brightness, the hottest and brightest being of the blue O class. The coolest include the reddish M- and S-type stars. To a first approximation, the O through M stars all contain very similar chemical elements but derive their different spectra through their different temperatures.

In the most modern system of naming, numbers are used after the letters to signify additional information. For example, the yellow supergiant star Mirfak (also known as Alpha Persei) is classified as F5 and is 5,000 times the luminosity of the Sun. The "5" indicates the star falls midway between F stars and G stars.

When observed using spectroscopes (instruments for forming and examining optical spectra) the Type-O stars show prominent hydrogen and helium emission and few or no emission lines of the other elements. The next hottest stars, the Type-B stars, show strong helium emission line spectra. Type-A stars are the first to exhibit noticeable spectral lines of metals like calcium, sodium, nickel, and iron. These emission lines grow in strength through classes F, G, and K, and serve as "fingerprints" for classes M, R, N, and S. The R- and N-type stars are often referred to as carbon, or C-type stars. In fact, the K, M, R, and N stars all show significant spectral features of carbon compounds. The presence of these carbon compounds tends to absorb the blue portion of the spectrum, partly contributing to the distinctive red color of R- and N-type giants. S-type stars are usually giant or supergiant stars and closely resemble M stars. Recent studies indicate that large organic molecules evolve within a few thousand years from chemicals in the gaseous envelope that surrounds some stars.[9] For example, astronomers have detected molecules of acetylene (C_2H_2, a building block for many more complex organic molecules) around stars. Also, the amino acid glycine has been discovered in Sagittarius B2, the giant molecular cloud near the center of the Galaxy. Such chemicals would eventually be spewed into interstellar space and find their way to planets. In the 1990s, astronomers discovered that the hot shock wave produced by comets hitting Earth's atmosphere encourage chemical reactions in which hydrogen cyanide and acetylene produce chemical units called amine groups.[10] These groups are components of amino acids, which are the building blocks of life's proteins. Formic acid (the stinging chemical ants squirt to defend themselves) has also been found in interstellar clouds.

In this chapter I've mentioned U.S. astronomers Cecilia Payne-Gaposchkin and Annie J. Cannon. These are just two of numerous women who contributed immensely to the field of astronomy. Table 3.2 lists several more.

Recent surveys of the most influential women in astronomy still living around the year 2000 reveal the following names: Jocelyn Bell Burnell, Nancy Boggess, E. Margaret Burbidge, France Cordova, Sandra Faber, Margaret Geller, Margherita Hack, Eleanor Helin, Roberta Humphreys, Christine Jones, Catherine Pilachowski, Mercedes Richards, Sally Ride, Nancy Roman, Vera Rubin, Carolyn Shoemaker, Jill Tarter, Jacqueline Van Gorkom, and Sidney Wolff. Just for fun, how many of these women can you identify or find on the web?[11]

I'm a particular fan of Cecilia Payne-Gaposchkin, who some have called the greatest woman astronomer of all time. Born in 1900 in Buckinghamshire, England, she came to the United States in 1923 to begin a 40-year career at Harvard College Observatory. A botanist aunt inspired her to be a scientist, but she experienced difficulties getting a science education and academic positions because she was a woman. Her overwhelming love for astronomy was her main driving force. Today we remember her for determining a temperature scale for

Table 3.2
Early Influential Women in Astronomy

Astronomer	Accomplishment
Caroline Herschel (1750–1848)	Various astronomical calculations; discovery of several nebulae and comets.
Maria Mitchell (1818–1889)	First professional American woman astronomer. Research on comet orbits and sunspots.
Annie Jump Cannon (1863–1941)	American astronomer who classified stellar spectra. Discovered hundreds of variable stars and five novae.
Henrietta S. Leavitt (1868–1921)	American astronomer who discovered the relationship between period and luminosity in pulsating stars.
Cecilia Payne-Gaposchkin (1900–1979)	First person to receive a Ph.D. in astronomy from Harvard. Spectral studies regarding chemical composition of stars.
Beatrice M. Tinsley (1941–1982)	Galactic evolution

the different types of stars. Her dissertation was titled, "Stellar Atmospheres, A Contribution to the Observational Study of High Temperature in the Reversing Layers of Stars," and it showed that the large variation in stellar absorption lines was due to differing amounts of ionization caused by stellar temperature and not to different abundances of elements. She correctly suggested that silicon, carbon, and other common metals seen in the Sun were found in about the same relative amounts as on Earth, but that helium and hydrogen were vastly more abundant. (Her discovery that hydrogen is the most abundant element in stars also meant that the Universe is mostly hydrogen.) Because her results disagreed with earlier theories, several contemporary astronomers said her results were "clearly impossible." However, most astronomers soon agreed with her that hydrogen was far more abundant in the Sun than in the Earth.

In 1977 she received the prestigious Henry Norris Russell Prize from the American Astronomical Society. In her acceptance speech, she discussed her excitement for science:

The reward of the young scientist is the emotional thrill of being the first person in the history of the world to see something or to understand something. Nothing can compare with that experience; it engenders what Thomas Huxley called the Divine Dipsomania. The reward of the old scientist is the sense of having seen a vague sketch grow into a masterly landscape. Not a finished picture, of course; a picture that is still growing in scope and detail with the application of new techniques and new skills. The old scientist cannot claim that the masterpiece is his own work. He may have roughed out part of the design, laid on a few strokes, but he has learned to accept the discoveries of others with the same delight that he experienced his own when he was young.[12]

In addition to her pure research, she was a skilled writer. Late in life, she wrote a fascinating tale of her life as an astronomer called "The Dyer's Hand." Here are some wonderful quotes from "The Dyer's Hand" reprinted in *Cecilia Payne-Gaposchkin: An Autobiography and Other Recollections*, edited by Katherine Haramundanis:

*   When I won a coveted prize . . . I was asked what book I would choose to receive. It was considered proper to select Milton, or Shakespeare. . . . I said I wanted a textbook on fungi. I was deaf to all expostulation: that was what I wanted, and in the end I got it, elegantly bound in leather as befitted a literary giant.
*   All motion, I had learned, was relative. Suddenly, as I was walking down a London street, I asked myself: "relative to what?" The solid ground failed beneath my feet. With the familiar leaping of the heart I had my first sense of the Cosmos.
*   At a very early age . . . I made up my mind to do research, and was seized with panic at the thought that everything might be found out before I was old enough to begin!
*   I have come to wish that all scientific work could be published anonymously, to stand or fall by its intrinsic worth. But this is an unrealistic wish, and I know it.
*   On the material side, being a woman has been a great disadvantage. It is a tale of low salary, lack of status, slow advancement. . . . It has been a case of survival, not of the fittest, but of the most doggedly persistent.
*   Do not undertake a scientific career in quest of fame or money. There are easier and better ways to reach them. Undertake it only if nothing else will satisfy you; for nothing else is probably what you will receive. Your reward will be the widening of the horizon as you climb. And if you achieve that reward you will ask no other.
*   There is no joy more intense than that of coming upon a fact that cannot be understood in terms of currently accepted ideas.
*   Nature has always had a trick of surprising us, and she will continue to surprise us. But she has never let us down yet. We can go forward with confidence—knowing that Nature never did betray the heart that loved her.[13]

British physicist Jocelyn Bell Burnell notes that Payne–Gaposchkin saw modern science evolve like a blossoming flower during her long career. Payne–Gaposchkin started her work before nuclear physics was mature and before the nature of galaxies outside the Milky Way was truly understood. By the end of her life she was studying infrared and ultraviolet data from satellites. Despite horrible prejudice because of her gender, she persevered and laid a trail for all women scientists to follow.[14]

***

As far back as 1917, before anyone really knew much about galaxies, researchers noticed that most galaxies had their spectral lines Doppler-shifted to the red, which suggested they were moving away from us at high speeds. The more distant the galaxy, the faster it moved. This is one of the main lines of evidence for the Big Bang that created the Universe and the subsequent expansion of space. However, it is important to remember that galaxies are not like flying pieces from a bomb that has just been exploded. Space itself is expanding. The distances between galaxies are increasing in the same way that black dots painted on a balloon's surface move away from one another when the balloon is inflated. It doesn't matter on which dot you reside. Looking out from any dot, the other dots appear to be receding. Note that, technically speaking, this recession sometimes only applies to clusters of galaxies rather than galaxies themselves, because those galaxies within a group are bound together by gravity. Our local cluster of galaxies, for example, is not expanding, but our cluster is moving further away from other clusters. One of the most important numbers in cosmology is called the Hubble constant (H_0). It measures the expansion rate of the Universe.

I don't really like driving in snow. There's something about the motion of the falling snowflakes that hurts my eyes, throwing my sense of balance all to hell. It's like tumbling into a field of stars.

— Neil Gaiman,
"Sandman #51:
A Tale of Two Cities"

Omnia quia sunt, lumina sunt. All things that are, are lights.

— Scotus Erigena,
Eleventh Century

CHAPTER 4

Luminosity and the Distance Modulus

> *Life is a narrow vale between the cold and barren peaks of*
> *two eternities. We strive in vain to look beyond the heights.*
> *We cry aloud, and the only answer is the echo of our wailing*
> *cry. From the voiceless lips of the unreplying dead there comes*
> *no word; but in the night of death hope sees a star, and*
> *listening love can hear the rustle of a wing.*
>
> — Robert G. Ingersoll, *The Ghosts and Other Lectures,* 1881

"**M**iss Muxdröözol! Open up!"

Bob bangs on the carbon doors that separate the museum corridor from Miss Muxdröözol's living quarters. Humanity entered the Carbon Age years ago, after soccer ball shaped cages of carbon atoms led to extraordinary new materials. The new materials are a hundred times stronger than any metal and weigh less than ordinary plastics. Unfortunately for Bob, they also block sound very efficiently.

"Miss Muxdröözol, where are you?" Bob looks down both ends of the dim corridor with his two faces. He knocks harder. *Please God, let her be here.* "Miss Muxdröözol!" Bob suddenly hears a sound coming from the direction of his living quarters.

A few seconds later, Miss Muxdröözol opens her door. "Bob, hi, what's up?" She has a cup of coffee in one hand and a blow dryer in the other. Currently, her hair is almost shoulder length, fashionably cut, with wonderful crimson streaks in it. A few strands of hair wriggle as if they have a life of their own, more like liquid solder than anything biological. Her cheekbones are high and pronounced, her mouth tenderly curved, her eyes skillfully shadowed. The only imperfection she has (though Bob thinks it only adds to her allure) is a Ψ-shaped scar near her left eyebrow—a memento of a childhood accident. She seems self-conscious about it.

"I need your help!" Bob says. "There's something in my bedroom." Bob steps into her apartment. On the mantel above her fireplace, below several swimming medals, is a small photograph of Hreithmar, her beloved grandmother. Her eyes are large red ovals with a splash of green. Miss Muxdröözol once told Bob she felt closer to her grandmother than her mother who never trusted or helped her.

Bob spins around so his multiple faces can get a full exposure to Miss Muxdröözol's quarters. The room looks how you'd expect an apartment of an unmarried theropod in her early 30s to look—a dump, with items of clothing draped over long-deactivated androids, take-out food boxes piled high on an ancient TV screen, and piles of holographic images of the Ooodles, the hot new singing band created from genetically engineered crabs.

On a low wooden desk is a collection of furculas, V-shaped bones formed by the fusion of the clavicles at the breastbone. Apparently it is customary for Miss Muxdröözol and her kin to collect the bones of their long-dead ancestors.

On her walls are pictures of her great-great grandparents, showing in explicit detail how their livers subdivide the visceral cavity into distinct anterior pleuropericardial and posterior abdominal regions. The diaphragmatic musculature is attached to a posterior colon.

Miss Muxdröözol walks closer to Bob, still holding her coffee cup. Her cheekbones are perfect, her jawline strong and feminine. On the couch, Bob notices the black beret she habitually wore along with her bioluminescent earrings.

How can someone so alluring be so alien at the same time? Bob is aware that her body is very different from his. Miss Muxdröözol has a trachea similar to those of crocodilians, situated ventral to the vertebral column. The rumor mill has it that under ultraviolet light, her liver fluoresces as a suboval, indigo-colored mass that extends from the vertebral column to the ventral body wall. Hepatic piston, diaphragm-assisted breathing in her species is powered by the diaphragmatic muscles that originate on the gastralia and insert on the lateral surfaces of the liver. As in other theropods, her triradiate pelvis is particularly well-suited to accommodate diaphragmatic muscle function.[1]

"Coffee?" she says, probably in an attempt to calm him down. She spills a drop as she offers him her cup, and a microcat darts out from beneath the couch to lap up the liquid.

Miss Muxdröözol's Microcat

Bob shakes his head. "Um, I found something weird in my apartment. Could—could you look? Maybe I'm hallucinating, but—"

There is a shuffling sound from beneath the couch as Mr. Plex scrambles out of an aestivation chamber, evidently fully refreshed.

"Mr Plex!" Bob yells. "What are you doing in Miss Muxdröözol's aestivation chamber?"

"No need to be jealous, Sir," Mr. Plex says as he towels off the remaining traces of phenyl tertiary butyl nitrone and warm cocktails of free radical scavengers. "I just needed a quick nap before our next lecture on stars."

Miss Muxdröözol comes closer. "Bob, no need to worry. You're under a lot of stress managing this museum gallery. Teach us more about stars. It will get your mind off your bad dreams and visions."

As Bob gazes into Miss Muxdröözol's azure eyes, he cannot help but think she is right. Could the creature in his room have been a fragment of a dream? Bob has heard that the ship's food supply may have been temporarily infused with a low-dose hallucinogen. It was a mix-up. One of the alien races required the chemical for sustenance, and the cook had accidentally mixed it into the human supply.

Bob needs to distract himself from the pressures. "Okay, but both of you must promise to check out my room when we're finished here."

"Certainly, Sir." Mr. Plex says as his mouth opens and closes spasmodically.

Bob goes to the window. "Look at the stars, Mr. Plex. What do you see."

"Stars. Some are dim. Some are bright."

"The brightness of a star that you see when you look at the night sky is its *apparent brightness*. But in actuality, you really can't tell for sure how much light the actual star shines into space by looking at it. After all, if you consider two equally bright stars at different distances from your eye, the closer one will be brighter than the distant one. A star's *luminosity* refers to the actual amount of light it emits each second, independent of its distance from us."

"Your Sun seems awfully bright," Miss Muxdröözol says as she strokes her microcat.

Bob sits down upon Miss Muxdröözol's furry sofa. "Astronomers are in the habit of comparing the luminosity of stars to the Sun's luminosity, symbolized by L_0. The Sun produces 3.86×10^{26} watts of power (or 5.18×10^{23} horsepower)."[2]

Mr. Plex jumps back. "That's a hell of a lot of horses. By the way, what's a horse?"

Bob waves his hand. "Never mind that. A watt is a unit of power equal to one joule of work performed per second, or to 0.0013404 horsepower." Bob speaks into a flexscreen located within the arm of the sofa. "Brunhilde, display power table." The following table appears:

Table 4.1
Power in Watts of Various Processes[3]

Process	Power
Solar power in all directions	10^{26}
Solar power incident on earth	10^{17}
Solar power incident on U.S. (average)	10^{15}
Solar power consumed in photosynthesis	10^{14}
U.S. power consumption rate	10^{13}
U.S. electrical power	10^{12}
Large electrical generating plant	10^9
Automobile at 40 mph	10^5
Solar power on roof of U.S. home	10^4
U.S. citizen consumption rate	10^4
Electric stove	10^4
One light bulb	10^2
Food consumption rate per capita	10^2
Electric razor	10^1

Mr. Plex looks at the last line of the table and then at the top line of the table. "A Sun is a hell of a lot of electric razors."

"You need not use that word 'hell' so often. But you are correct. The Sun's power is equivalent to 38,600 billion trillion electric razors turned on all at once. And the most luminous stars are over a million times more luminous than the Sun. The dimmest stars known are ten thousand times dimmer than the Sun. However, remember that a star's *apparent* brightness to us on Earth depends on both its distance from us and its luminosity. Light spreads out in all directions, and, in particular, the amount of starlight falls off as the square of the distance. This means that if two stars have exactly the same luminosity but one is three times as far away from us as the other, the distant star will look only $(1/3)^2$ or 1/9 as bright as the closer one."

Miss Muxdröözol nods, "Yes, if you move a light twice as far away, it appears four times fainter."

"We can make this very obvious with an example using cats. Miss Mux-dröözol, may I borrow your microcat?"

She backs away. "There's no way your using him in your experiment."

Bob puts up his hands. "Okay, okay. We'll simulate this on the flexscreen using a toy cat. Consider two circles, one with a one-foot radius and the other with a two-foot radius." Brunhilde displays figure 4.1. "Place a toy cat along

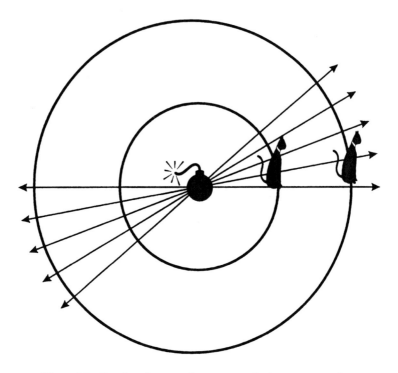

Figure 4.1   Bomb and toy cat demonstrate the inverse square law.

each circle. Place a bomb in the center of the circles, and let it explode. The bomb fragments are like photons of light."

Bob points to the flexscreen. "You see five arrows that indicate five bomb fragments flying outward. The area of the outer circle is four times the inner circle because the area of a circle is $4\pi r^2$. Now, look at the diagram. The cat in the inner circle is hit by four fragments. Because there is more area to cover, the cat on the outer circle is hit by only one fragment. If this experiment were performed with photons of light and detectors, four times more photons would strike the inner detector than the outer one. And this is why starlight falls off as the square of the distance."[4]

The fur on the couch begins to wave slightly as Mr. Plex lowers his abdomen onto the couch. "Sir, is there a way to accurately compare how bright different stars appear to us?"

"Yes, the *apparent magnitude* measures how bright a star appears. The ancient Greek astronomer Hipparchus (who died around 127 B.C.) classified the stars he saw into six categories, or magnitudes, by their relative brightness. Today, astronomers use photometers to measure brightness. The brighter the star, the more negative its apparent magnitude."

"Sir, can you give examples?"

"The sun has an apparent magnitude of −26.7. Alpha Centauri, the closest star in the night sky, has a magnitude of 0. A full moon is somewhere in between with an apparent magnitude of −12.5." Bob speaks into the flexscreen. "Brunhilde, show magnitudes." The following table appears on the screen.

Table 4.2
The Five Brightest Stars in the Night Sky

Star	Apparent Magnitude	Absolute Magnitude	Spectral Class	Distance (light-year)
Sirius	−1.46	1.4	A	8.6
Canopus	−0.72	−2.5	A	74
Alpha Centauri	−0.27	4.1	G	4.3
Arcturus	−0.04	0.2	K	34
Vega	0.03	0.6	A	25

"A difference of 1 magnitude means a brightness ratio of about 2.5. That means we receive about 2.5 times as much light from a star of magnitude 0 as we do from a star of magnitude 1."

"I don't get that last point," says Miss Muxdröözol.

"Astronomers often use a *magnitude scale* when comparing stars. It's a very nonlinear scale where a brightness ratio of 100:1 corresponds to a difference in magnitude of 5, but a brightness ratio of 251:1 corresponds to a difference in magnitude of only 6. It's so nonlinear because of our visual systems. What we perceive as a linear increase in brightness of one magnitude corresponds to a real increase in brightness of the fifth root of 100 or 2.511886. Using this scale, a first-magnitude star is 100 times brighter than a sixth-magnitude star."

Bob takes some mascara from Miss Muxdröözol's pocketbook and scrawls an equation on her walls.

$$r = 2.511886^{\Delta m}$$

"A formula describes the relationship between difference in magnitude Δm and brightness ratio r. Try plugging numbers in, and you will be amazed."

Miss Muxdröözol continues to stroke her microcat, which begins to purr. The purr seems vaguely alien—slightly higher in pitch than for a natural cat. "Bob, the apparent magnitude seems to tell us little about the actual brightness of a star."

"You're right. What astronomers often care about is the *absolute magnitude*, a measure of the luminosity or how much light a star really shines into space. A star's absolute magnitude is the apparent magnitude the star has if it were located ten parsecs from us. You can conceptualize this by lining up all the stars of heaven at the same distance from Earth. This eliminates the distance effect that alters how bright the stars appear to us on Earth" (figure 4.2).

"Sir, does this mean that if a star is farther than ten parsecs from us, the value of its apparent magnitude is larger than its absolute magnitude."

"Yes, recall that large positive magnitudes indicate faint objects. Consider the star Alnilam. It's 460 parsecs away, and its apparent magnitude is 1.7 whereas its absolute magnitude is –6.6. Also, if a star is closer than 10 parsecs, the value of its apparent magnitude is smaller than its absolute magnitude. For example, the star Altair is 5.14 parsecs away. Its apparent magnitude is 0.77 and its absolute magnitude is 2.22."

Miss Muxdröözol studies the previous flexscreen table showing absolute and apparent magnitudes of the brightest stars. "Let me see if I have this right. Alpha Centauri looks brighter than Arcturus because its *apparent* magnitude is smaller than Arcturus's. However, Arcturus is really more luminous than Alpha

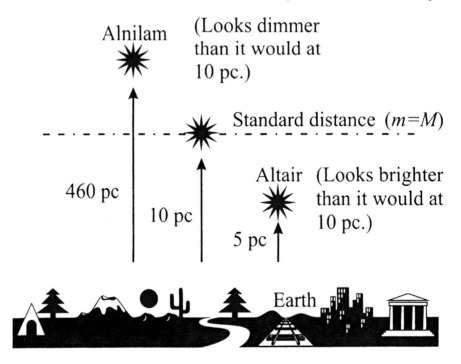

Figure 4.2   A star's absolute magnitude (*M*) is the apparent magnitude (*m*) of the star if it were located ten parsecs from us.

Centauri because its *absolute* magnitude is smaller than Alpha Centauri's. The reason for this 'reversal' is that Arcturus is so much further away that it appears dimmer to our eye than Alpha Centauri."

"Superb! Now the last thing I want to teach you about today is the *distance modulus*." Bob says the phrase "distance modulus" slowly so as to heighten his friends' suspense and increase their appreciation of his scientific prowess.

Mr. Plex lets out a great gust of wind. It smells like the decaying flesh of some long dead nematode at the height of its estrous cycle.

"The distance modulus is the difference between the apparent magnitude (*m*) and the absolute magnitude (*M*)." Bob takes out a business card from his pocket and hands it to Miss Muxdröözol. The card reads:

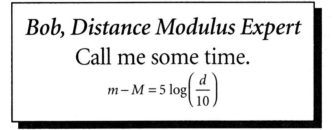

Bob, Distance Modulus Expert
Call me some time.
$$m - M = 5 \log\left(\frac{d}{10}\right)$$

"Here *d* is the distance from Earth in parsecs, and the distance modulus is simply *m-M*. Sometimes you see this expressed in a related way: $M = m + 5 - 5 \log d$. Astronomers can measure the apparent magnitude *m*. Sometimes we can estimate a star's absolute magnitude *M* from its stellar spectrum, and therefore we can compute the star's distance from us. This can be useful if we want to estimate how far away a distant star is. Remember that stellar parallax is difficult to use when measuring distances to faraway stars."

"Sir, this means that if you give us values for apparent and absolute magnitude, spectral class, or parallax, there's so much we can tell you about the star."

"Yes. Let's review. The *spectral class* tells us about the star's temperature. The *apparent magnitude* tells us how bright a star looks. The *absolute magnitude* tells us how luminous a star actually is. *Parallax* tells us how far away a star is."

Bob looks back and forth between Mr. Plex and Miss Muxdröözol. "Miss Muxdröözol, could you take a look in my living quarters? You've got to tell me if what I'm seeing is a hallucination."

She laughs, "Bob, do I look like a shrink to you?" When Bob doesn't answer, she smiles a smile that transforms her face into pure sunlight. "Is this some sort of line to get me back into your room?" Her fingers play with an ameba-shaped pendant that hangs from a chain around her neck.

"This is serious," he says as his gaze drifts to her walls that contain a history of science diploma from Harvard with photos of her standing next to android versions of Bill Clinton, Jimmy Carter, and Charlton Heston. No wonder she knows so much about a broad range of subjects.

Her smile fades. "Okay, let's all go. What's this about?"

Bob laughs nervously with one face and then the other, "Both of you. Just come with me. It'll only take a second. I—I need someone to tell me if the thing's real. I know it sounds crazy. Just come."

Mr. Plex backs up, "Sir, are you sure —"

Bob rubs his hands together. "Please." They leave her quarters, and Bob speaks into his flexdoor, "Brunhilde, open." Bob slowly enters his apartment. "The thing is in the bedroom."

"You're making me nervous," Miss Muxdröözol says. She and Mr. Plex gaze down at the colorful Miro lithograph that has fallen to the floor. With a sudden stop, she stands riveted and faces Bob, "You sure about this?"

"C'mon, we won't get too close. It didn't seem to move much."

Her corneas widened. "It? What are you so afraid of, a mouse?"

The swishing sounds are no longer coming from behind the bedroom door, and Bob can see the door is still closed. *It must still be in there.*

"What's that smell?" Miss Muxdröözol says putting her hand up to her triple nostrils. "Phew, it smells like dead fish. One of the fish from your tank jump out?"

"C'mon."

When they reach the bedroom door, Bob turns to Mr. Plex and says, "Tell me what you see." Bob reaches out with a trembling hand and pulls the knob.

"Nothing," Mr. Plex says as he manually moves his dorsal ommatidia in the direction of the room. Several of his pseudopupils wander chaotically as if trying to observe his surroundings from several vantage points. One of the pseudopupils actually pops from his head as Mr. Plex strains and stretches his visual system to the limit.

"Nothing," Miss Muxdröözol says. Now there is a trace of irritation in her voice.

"Wait," says Bob as he cautiously pushes the door open a few more inches, cringing as its biohinges make a squeaking sound. *Where is that creature?* He pushes the door open a little further, and it squeaks like a big rat. *Where is it?*

"Let me," Miss Muxdröözol says impatiently, shoving the door open wide. She looks over Bob's shoulders, "I don't see anything."

While still staying outside the room, Bob slowly reaches into the room and begins to grope for the piezoelectric light switch on the wall.

"Where is that damn switch?" he says through clenched teeth. Gaining control over his shattered nerves, Bob quickly reaches into the bedroom and sweeps his fingers upward to catch the switch that should be there. Finally he finds it, turns on the ceiling light, and quickly withdraws his hand from the bedroom.

Miss Muxdröözol steps forward. "You act like something's going to bite you."

"Miss Muxdröözol, stay behind me!" Bob quickly shifts into a graceful Tai Chi motion called "Grasping the Sparrow's Tail," his ever-moving hands producing a wall of muscle through which past aggressors have never dared transgress.

Miss Muxdröözol takes a hasty step backward.

Bob looks to the corner of the room and relaxes his defensive posture. "It *was* right there."

"What are you talking about?" They all step into the bedroom. "Look, I already told you—if this is some scheme to get me in your bedroom—"

Miss Muxdröözol stops and sees a movement by the head of the bed. "Oh no," she says, tosses back her head, and screams. Bob puts his fist in one of his mouths and steps back.

The veined membrane is undulating near the bed. The two pieces he cleaved evidently reformed into one being, and there is no sign of the small naked brain. The creature's sheer physical smoothness is alien, intimidating.

"It can't be from this museum ship," Mr. Plex mumbles as if he is drunk. Miss Muxdröözol opens her mouth but nothing comes out.

"Sir, let's get out of here," Mr. Plex says grabbing Miss Muxdröözol's vestigial thumb. Bob nods, and they begin to close the bedroom door when some of the veins start changing colors. They oscillate red, white, green, and blue, more beautiful than any Christmas-light display Bob has ever seen.

"Look," Bob says, pointing.

"What do you mean, 'look'?" Miss Muxdröözol says, squeezing Mr. Plex's claws even tighter. "Let's not look. Let's get out of here."

But she too is drawn to the gorgeous rainbow before them. The flashing colors are hypnotic and turn the alien creature into a thing of captivating beauty.

Bob and his friends are still, hypnotized by the membrane that expands to their size in a few seconds and then suddenly sucks them inside a portal. For a minute all the bones in Bob's body seem to vibrate, and he feels as if he is surfing on a large ocean wave. The wave begins to lift them higher and higher, and when they reach the crest, they begin to glide out into space.

They soon become surrounded by triangles and lines that seem to extend outward to angular infinity. Some of the colors are so bright that for an instant Bob thinks his eyes will burn away, but soon the intensity of light dims to acceptable levels.

Bob is in a translucent funnel into which pours colored "oil" swimming with geometrical shapes and flowers. The fabric of reality is tearing, like a fragile piece of cloth. Another world pours through the funnel—comes rushing in. Bob sees an afterimage of his bedroom that glows for a few seconds and then vanishes.

"Bob!" Miss Muxdröözol screams.

"Keep holding my hand." He is mesmerized by the marvelous ballet taking place in his head.

"Don't let go of me." Miss Muxdröözol's hand trembles.

Bob looks around for Mr. Plex but cannot find him. Bob feels as if he is trapped in a narrow enclosed space. The air is chilly for a second but then it becomes warm and comfortable, almost as if their invisible enclosure is upholstered in some kind of sensual material. *Oh God*, Bob thinks as fear flutters around his brain like a flock of nervous butterflies. There is a burst of light and then a screaming symphony of birds.

The lights are fading. *Everything's turning black*, Bob thinks. Black. Black. Black. The touch of Miss Muxdröözol's vestigial thumb is the only real thing in an ever-changing reality.

Some Science Behind the Science Fiction

The greatest mystery is not that we have been flung at random between the profusion of matter and of the stars, but that within this prison we can draw from ourselves images powerful enough to deny our nothingness.

— Andre Malraux, *Man's Fate*, 1933

The luminosity L of a star provides astronomers with much information. For example, the lifetime τ of a star is proportional ($\propto$) to its energy supply divided by the rate at which it is used. Because energy supply is proportional to mass, and the rate is proportional to luminosity, we find that $\tau \propto M/L$. Similarly, because luminosity is proportional to the mass to the 3.5 power ($L \propto M^{3.5}$, an average relation described by James Kaler)[5] we can approximate stellar lifetime by $\tau \propto (1/M)^{2.5}$. This means, for example, that Vega, a typical A-type dwarf star, which is 2.5 times the mass of our Sun will live $(1/2.5)^{2.5}=1/10$ as long as the Sun. There are stars in our Universe that only survive for a few million years while others that are less than 0.8 solar masses have lifetimes longer the age of the Galaxy.

* * *

One of the classic problems of astronomy is "Why is the sky dark at night?"[6] This question was asked by the astronomer Johannes Kepler in the early 1600s, when the largeness of the heavens was starting to be fully appreciated. In 1832, the German astronomer Heinrich Wilhelm Olbers presented a paper that discussed this issue, and the problem subsequently became known as *Olbers' paradox*. (Note the spelling is "Olbers" not "Obers" or "Olber;" this is one of the most commonly misspelled names in all of science!) Here is the problem. If the Universe is infinite, as you follow a line of sight in *any* direction, that line must eventually intercept a star. This implies that the night sky should be dazzlingly bright with light coming from stars. You might retort callously that "the stars are far away and that their light dissipates as it travels such great distances." Sounds nice, but you are naive. This is *not* why the sky is dark. Star light dims as it travels, but by the square of the distance from the observer. However, the *volume* of the Universe and hence the total number of stars would grow as the cube of the distance. So, even though the stars become dimmer the further away they are, this dimming is compensated by the increased number of stars. This reasoning is schematically illustrated in figure 4.3.[7] If we lived in an infinite Universe, the night sky should indeed be very bright.

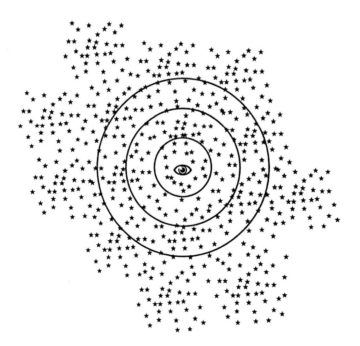

Figure 4.3   Prelude to Olbers' paradox. If we assume a constant number of
stars per unit volume of space, the number of stars in each spherical shell increases
with distance from your eye. This increase compensates exactly for the fact that
more distant stars are further away and individually would appear fainter to us.
Each shell of stars contributes the same amount of light to the night sky.

Here's the solution to Olbers' paradox: We do not live in an infinite and static Universe. The Universe has a finite age and a non-Euclidean spacetime geometry; that is, it is expanding. Because only around 12 billion years have elapsed since the Big Bang, we can only observe stars out to a finite distance of about 12 billion light years. This means that the number of stars, about ten thousand billion billion (10^{22}), is finite. Because of the speed of light, there are portions of the Universe we never see. Because light travels at a known, constant velocity, light from very distant stars has not had time to reach the Earth. The first person to suggest this resolution to Olbers' Paradox was the writer Edgar Allan Poe. (In his prose poem "Eureka" [1848], Poe discusses the paradox as well as other scientific ideas.)

Another factor to consider is that the Universe is expanding. This expansion acts to darken the night sky because the distant stars brighten the night sky less than they would in a simple Euclidean space. Starlight expands into an ever vaster space. Also, the Doppler effect causes a redshift in the wavelengths of light emitted from the rapidly receding stars.

Imagine what would happen if our Universe had an infinite age or was static. The night sky would be as bright as a star's surface, and life would not evolve. The planets would be fried. Religious people may find this interesting. They might argue that we would need God to *create* a Universe (i.e., the Universe could not have persisted for all eternity) for life to evolve using the known laws of physics. We "need" a Universe that is finite in space and in time. There has to be a *beginning* or reality would become extinct.[8]

> "Space isn't remote at all. It's only an hour's drive away if your car could go straight upwards."
>
> — Sir Fred Hoyle

CHAPTER 5

Hertzsprung-Russell, Mass-Luminosity Relations, and Binary Stars

Men can do nothing without the make-believe of a beginning. Even Science, the strict measurer, is obliged to start with a make-believe unit, and must fix on a point in the stars' unceasing journey when his sidereal clock shall pretend that time is Nought.

— George Eliot (1819–1880), *Daniel Deronda*

The veined membrane is nowhere in sight. We can be thankful, at least, for that.

Bob, Mr. Plex, and Miss Muxdröözol are holding hands as they find themselves standing in an endless field of grass. The grass is violet and two to four feet high. The *Picasso*, Bob's museum ship, is gone. It is dusk but the grass appears to bioluminesce, making it easy for Bob to see Mr. Plex and Miss Muxdröözol. In the sky is a bright galaxy with a dense spherical core surrounded by several spiral arms.

Bob looks down. The ground is flat. The air is fresh with the faint ozonic smell of rain. Bob is momentarily paralyzed with the shock of suddenly being in a new environment.

"Where the hell are we?" Miss Muxdröözol whispers. She looks at Bob as if waiting for soothing words, some explanation for their strange predicament. "Are you all right?"

Bob touches the ground beneath his feet to see exactly what he is standing on, but there is nothing unusual about the soil, as far as he can tell. He feels lighter. Gravity seems much lower on this world than on Earth.

Mr. Plex comes closer. "Sir, I know exactly where we are and how we got here."

"What?"

"We are on an asteroid. That veined membrane is a member of a race called the Valkyries. They have used their wormhole technologies to transport us here.

This asteroid is 30 thousand light years perpendicular to the Milky Way's galactic plane. Look up. That's our Milky Way above our heads." Mr. Plex removes a pocketscreen from a crevice in his body and examines the device. "My gravimeter program suggests that this is indeed a typical asteroid, but the gravity strength is slightly down from what I expect."

Bob puts his hands on his hips. "Wait, how do you know all this?"

"As we were transported, the Valkyrie implanted a message in the diamagnetic crystals of my forebrain."

"I don't suppose it told you *why* it took us here?"

"No, but it did tell me that we are in a very large transparent enclosure on the asteroid's surface. There's an atmosphere in the enclosure to make us comfortable. I suppose the grass is to make us feel comfortable. Perhaps it's edible."

They begin to walk—sometimes hop—as Bob's eyes adjust to the dim light and the weak gravity. Now he sees that in one direction, beyond the field of grass, is a cratered surface. "Wait, I recognize this asteroid. It's the Nibelungenlied. But some of the craters are different. And look over there, that huge fault is not on the Nibelungenlied."

"It *will* be," says Miss Muxdröözol.

"What's that mean? What do you mean by 'will'?"

"It means that the wormhole creature not only transported us in space but also forward in time. That fault didn't occur in the past or our telescopes would have already revealed it or traces of it in the present. Hence, we are probably in the future."

Mr. Plex nods. "We are exactly 81 million years in the future. The thermal indicator on my pocketscreen scans to check the background radiation temperature of the Universe."

Miss Muxdröözol looks puzzled.

Bob nods. "The Universe is filled with the remnant heat from the Big Bang, the explosion that created our Universe somewhat over 10 billion years ago. The leftover heat we measure today is called the *cosmic microwave background radiation*."

"Is it hot?" asks Miss Muxdröözol.

"In the twenty-first century," Mr. Plex says, "this radiation was very cold: only 2.728 degrees above absolute zero. It fills the Universe and can be seen everywhere we look. However, right now, my device measures a slight decrease in this radiation, which leads me to believe that our strange journey plunged us 81 million years into the future."[1]

"Eighty-one million years," Bob says. "My God. That's about the same amount of time between the age of dinosaurs and the first appearance of humans on Earth."

Miss Muxdröözol lets out a low whistle. "I wonder what has happened to Earth in 81 million years. What have humans evolved into by now? What does my species looks like? And Mr. Plex's?"

Bob looks up. "Stars are still being born now but more in the spiral arms than in the core. The core stars formed early in the Galaxy's history and are late in their evolution. The exploding stars, or supernovas, in the arms are kicking out heavy elements like carbon that were manufactured in the stars."

"Are the stars really different in the core of the Galaxy than in the spiral arms?" asks Miss Muxdröözol.

Bob takes the pocketscreen from Mr. Plex and says, "Brunhilde, display Population I and II stars." Bob pauses. "Miss Muxdröözol, take a look at this table."

Table 5.1
Stars, Young and Old

Class	Location	Age	Temperature	Elements
Population I	Spiral arms	Young	Luminous (e.g. our Sun) and hotter than Population II stars	Rich in heavy elements
Population II	Galactic center	Old	Red, cool	Poor in heavy elements

"In the 1940s, astronomer Walter Baade determined that the Galaxy had two different classes of stars, which he called *Population I* and *Population II* stars. Most of the Population I stars are like our Sun and found in the Milky Way's spiral arms that coil around the galactic center. Population II stars are red and cool and found in the center of the Galaxy. Population I stars are richer in heavy elements than Population II stars because Population I stars are young, born in the stellar nurseries of the galactic arms. These baby stars have had a chance to feed off the remnants of their predecessors that exploded or gently puffed off their heavy elements before they died. Each successive generation of stars is richer in heavy elements than its dead parents that cooked up heavy elements and scattered them to the interstellar gases and dust from which the babies are born."

"Sir, the veined Valkyrie said something else during our transport here," says Mr. Plex. "It wants you to continue teaching us about stars, even here."

"Why?"

"I'm not sure. Perhaps to prepare us for something—something big."

"Like what?"

"Didn't say."

The three of them sit down in the field of waving grass. "Mr. Plex, a little while ago I told you about stellar luminosities and temperatures. In the early twentieth century, two astronomers discovered an interesting relationship between luminosity and stellar temperatures. Henry N. Russell (1877–1957) and Ejnar Hertzsprung (1873–1967) noticed that a plot of luminosity versus temperature give vast insight into stars. Today we call such a diagram a *Hertzsprung-Russell* (or *H-R*) diagram."

Miss Muxdröözol slaps her hands together. "Hey, shouldn't we be trying to figure out how to get off this asteroid and looking for food and water? Are we nuts talking about stars at a time like this?"

"Certainly," Bob says. "Brunhilde, display H-R diagram." Brunhilde displays figure 5.1. Bob turns to Miss Muxdröözol, "You can read the spectral class on the horizontal axis and luminosity and absolute magnitude on the vertical axis. I've indicated where a majority of stars seem to fall in the dark regions with little white spots. What's the first thing you notice?"

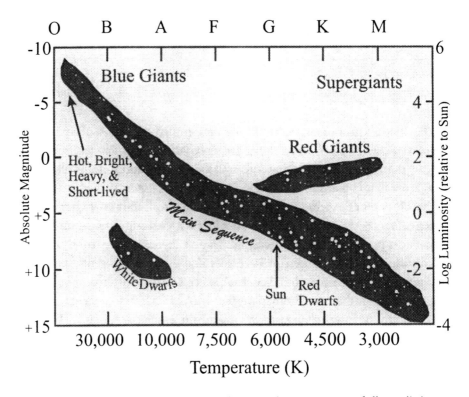

Figure 5.1   Hertzsprung-Russell (H-R) diagram. About 90 percent of all stars lie in a diagonal band called the Main Sequence.

"The stars seem to be clustered. I see a long diagonal cluster and a few other smaller clusters."

"Right! And that means that there is a significant relationship between a star's absolute magnitude and its temperature. Otherwise, the dots would be splattered all over the diagram like random sprinkles of ink. About 90 percent of the stars occur in the diagonal band called the *Main Sequence*. The upper left part of the Main Sequence band contains hot *blue stars*. Nearby are the very bright *blue giants*. The lower right part contains the cool, dim *red dwarfs*. Red dwarfs are the most common type of star near to Earth.[2] Most of the stars off the Main Sequence are the cool, luminous giants and supergiants or hot, dim white dwarfs. I'll explain what these stars are later when I talk to you about stellar evolution."

Brunhilde, of her own volition, displays figure 5.2, which shows the same kind of information as figure 5.1, with an emphasis on stars' luminosities in comparison to the Sun's luminosity.

The air is fresh, and Bob can hear the rustle of grass as they move through it. After a few minutes, they come upon a path and a cut lawn. There is a black bicycle on the path that leads to a house, but it is just a facade, a two-dimensional cutout. Next to the house is an ordinary-looking dogwood tree.

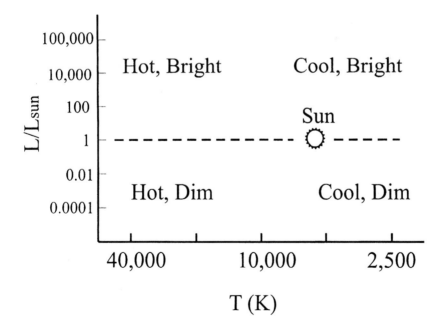

Figure 5.2   Temperature and luminosity. The *y*-axis expresses luminosity relative to the Sun's luminosity. See previous figure for additional details.

Mr. Plex begins taking digital pictures of the house using his photoreceptor chips embedded in his zygomatic arches. Next he sets up a secure satellite link to Earth. "Sir, hopefully, these images will bounce from the NSA/Australian Station at Pine Gap to the Vint Hill Station just north of Manassas, Virginia, and finally to the Pentagon. Let's tell them about some of this."

"Mr. Plex, you've forgotten we're millions of years in the future. Certainly, if humans are around, there's no Pine Gap, Vint Hill Station, or Pentagon! We're not even sure where the Earth is. I'd put that stuff away."

Bob looks at Miss Muxdröözol. "Now, it's time to talk about stellar *mass-luminosity relations*." Bob pauses a moment to see if the last phrase has caused Miss Muxdröözol to be impressed with his intellectual acumen. "First, I want to focus on the main sequence of the H-R diagram. The mass of a star can be guessed at pretty easily by looking at the Main Sequence. Brunhilde, show mass-luminosity relation." Figure 5.3 appears on Bob's flexscreen. "You can see that a star's position on the Main Sequence depends on its mass. At the bottom right are low-mass stars that are least luminous, and at the top left are high-mass stars that are most luminous. Notice that Brunhilde has indicated the masses of some stars along the Main Sequence in comparison to the Sun."

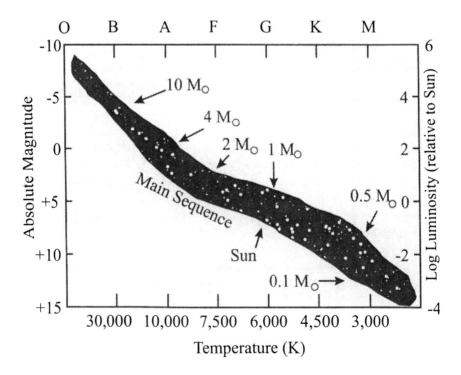

Figure 5.3   Stellar masses in the Main Sequence.

"Sir, what are we really learning here?" asks Miss Muxdröözol.

"Mass determines a star's luminosity and temperature. You can estimate a star's relative luminosity L_r (with respect to the Sun) from the star's relative mass M_r (with respect to the Sun) by $L_r = M_r^{3.5}$. Very small, cool stars would be below the chart. For example, a *brown dwarf* is an astronomical object with characteristics that are intermediate between a planet and a star—and it is 50 times smaller than our Sun. The biggest stable stars have masses about 70 times the Sun's mass."

"Sir, okay, we know that mass determines the position of a star in the Main Sequence of the H-R digram, but what can we know about a star's size?"

"Except for the Sun, stars are so far away that we really can't measure their size directly. The Sun's diameter is 1.4 times 10^9 meters (about 860,000 miles). That means you could put 109 Earths side by side along the diameter of the Sun. Here, this will give you an idea." Bob puts 109 small pebbles in a row.

109 Earths Fit Along the Diameter of the Sun

"If we know the temperature and luminosity of a star, then we can estimate the star's radius from the *Stefan-Boltzmann Radiation Law*." Bob hands Miss Muxdröözol a business card with an equation:

> ❧ I Like It Hot! ❧
> *Stefan-Boltzman Radiation Law*
> $$L = 4\pi R^2 \sigma T^4$$

Miss Muxdröözol points her vestigial finger at Bob. "Do you show that to all your lady friends?"

Bob decides to be silent on the matter. "In the equation, *L* is the luminosity, *R* is the star's radius, and *T* is the temperature. Sigma, represented by the Greek letter σ, is the Stefan-Boltzmann constant and equal to 5.67×10^{-8} watt per meter^2 K^{-4}. Don't worry about the details. The law merely states that the total energy emitted by a star is roughly proportional to the fourth power of its absolute temperature. A star that is twice as hot as our Sun (but with the same radius) radiates 16 times more energy than the Sun. You'll find that the radii of stars on the Main Sequence span in size from white giants, which are 25 times the Sun's radius, to red dwarfs, which are about a tenth the Sun's radius."

"Bob, I asked if you showed that card to all your lady friends?"

Mr. Plex ambles forward. "Sir," he says, "I only show cards like that to women with tracheas extending from the larynx's cricoid cartilage down to the division of the left and right mainstem bronchi at the fifth thorasic vertebra."

For a moment, Miss Muxdröözol is quiet. A wind blows through the grass, and then she says, "Get me out of this nuthouse."

"Sir, which stars are the biggest?" asks Mr. Plex.

"The supergiants. Take a look at the star Betelgeuse in the constellation Orion." Bob draws figure 5.4 on the dirt with a stick, and Brunhilde displays figure 5.5. "You can see the star's diameter is much larger than the diameter of Earth's orbit around the Sun. Indeed, Betelgeuse is so big that, if you stuck it at the center of our solar system, its outer atmosphere would extend past the orbit of Jupiter. This baby is huge."[3]

Bob draws lines indicating the diameter of Earth's and Jupiter's orbit. "Betelgeuse was so big that Andrea Dupree of the Harvard-Smithsonian Center for As-

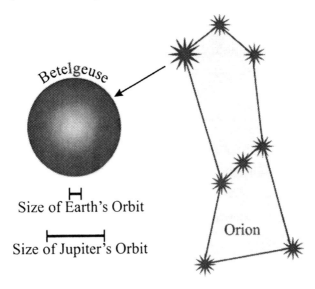

Figure 5.4   Betelgeuse, the second brightest star in the constellation Orion. Betelgeuse, a red supergiant, is one of the largest stars known and has a diameter around 500 times the diameter of the Sun.

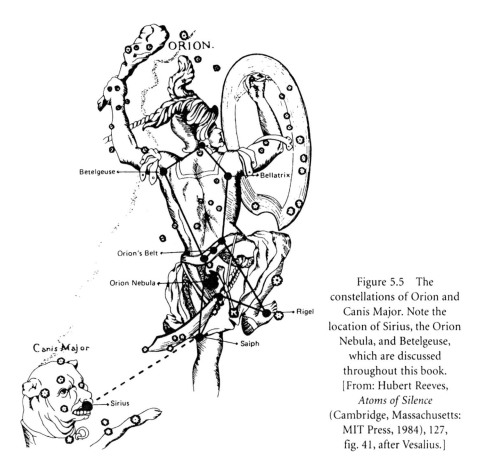

Figure 5.5   The constellations of Orion and Canis Major. Note the location of Sirius, the Orion Nebula, and Betelgeuse, which are discussed throughout this book. [From: Hubert Reeves, *Atoms of Silence* (Cambridge, Massachusetts: MIT Press, 1984), 127, fig. 41, after Vesalius.]

trophysics could obtain the first direct image of the surface of a star other than our Sun." Bob pronounces the star's name as BET"l-jooz. "Dupree's pictures, made with the Hubble Space Telescope in the late 1990s, showed a mysterious, hot, bright spot instead of the usual myriad of little sunspots that our own Sun exhibits. The big spot was 2,000 K degrees hotter than the surrounding surface of the star."

Bob walks over to the house's facade and taps on it with his knuckles. It seems to be made of wood.

He turns back to Miss Muxdröözol. "Betelgeuse is one of my favorite stars; it has a 420-day period, during which it oscillates in size. Think of it as ringing like a bell. By now the star must have exploded. It was always a prime candidate for imminent self-destruction in a supernova explosion."

Deep grumbling noises come from beneath the ground, and Bob thinks he hears the buzzing of helicopters. Hoping that his confusion will dissipate, he stands still for a few seconds. Suddenly, hundreds of white dogwood blossoms, stripped from the nearby tree by the wind, blow across his face.

Miss Muxdröözol comes closer to Bob. "This place is giving me the creeps."

When everything is quiet, Bob resumes his discussion, "Betelgeuse is a red giant superstar that could fit a million Suns inside of it. As comparison, white dwarf stars are about the size of Earth. We'll talk about these giant and tiny stars later. For now, just remember that, through time, a star leaves the Main Sequence. You can depict this in the H-R diagram by the star migrating off the diagonal region in which the Main Sequence resides. A star's core contracts and becomes hotter. Hydrogen burning ignites in a shell around the core, causing the envelope to expand, but as it does, it cools. The cooler, larger star becomes redder and more luminous. Eventually, the temperature in the core reaches 100 million degrees K and helium burning ignites in the core. At this point in the star's evolution, the star has a surface temperature around 3,500 degrees K and a radius equal to the orbit of Mars. In general, supergiants have diameters several hundred times that of the Sun and a luminosity nearly one million times as great. But the big stars don't have long lives—only a few million years, extremely short if you consider a star's overall lifetime."[4]

"Let's walk," says Bob. He stretches his sore legs and once again explores the peculiar surroundings. He looks left and right, searching for an object or aspect of the grassy field that he might recognize. The three of them walk for minutes, but to their chagrin and stupefaction, the grass continues to stretch in many directions for as far as his eyes can see. It's nothing but an endless plain of violet grass, strange varicolored weeds, and a few stunted trees.

"When we talk about black holes, I'll tell you more about the density of stars. For now, realize that the Sun is actually only slightly more dense than water. Because red giants have about the same mass as the Sun in a huge volume, their average densities are very low—about the same as that of vacuums produced on Earth. On the other hand, consider the white dwarfs. These are faint stars that are the evolutionary endpoints of intermediate- and low-mass stars. A majority of stars, like our Sun, end their lives as white dwarfs. If I were to give you a thimble full of white dwarf mass, you could never lift it. A teaspoon of white dwarf material weighs several tons on Earth. The time needed for a white dwarf to cool down is billions of years."

Occasionally Bob sees gases oozing from mist-covered swamps and the movement of reeds and cattails in some of the smaller marshes. In these algae-infested pools stand squidlike creatures of various sizes and colors. Everywhere is the smell of decay—the putrefaction of nearly liquid masses resembling the relucent remains of long-dead ruminants. By comparison, this stench made the stink of a cesspool seem like a new perfume by Chanel.

Miss Muxdröözol pinches her triple nostrils. "Uggh, let's stay away from them."

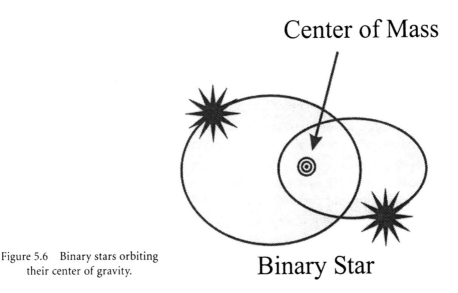

Figure 5.6   Binary stars orbiting their center of gravity.

Bob points to the pocketscreen. "Miss Muxdröözol, let's talk about *binary stars*. These are two stars that revolve around a common center of gravity. Numerous stars reside in binary systems" (figure 5.6).

Bob touches a star on the pocketcreen so that the star begins to blink, "Now, take a look at this nice star" (figure 5.7).

"The one in the handle of the Big Dipper?"

"Yes, it's Mizar in the constellation Ursa Major, and it's actually a *visual binary*. When you look at Mizar with a telescope it appears to be a pair of stars in

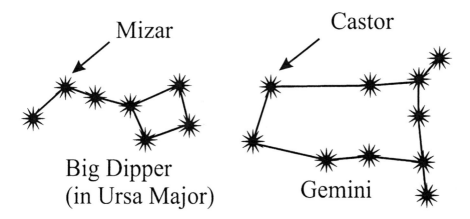

Figure 5.7   Italian astronomer Giovanni Battista Riccioli in 1650 found that Mizar in Ursa Major was a visual binary that appeared to consist of two stars. Today we know that the star is a quadruple star, which is even more complex. And Castor in the Gemini constellation is actually a collection of six stars in a crazy gravitational orgy.

orbit around each other. Half of all stars in the Milky Way Galaxy are binaries or members of more complicated couplings." Bob pauses and looks at Miss Muxdröözol. "Sometimes you can figure out the masses of the stars by their distance from us, angular size, and period of their orbits."

Bob arranges five pebbles on the ground. "There are five kinds of binary stars, and they are distinguished by how the stars are observed. If the two stars can be visually separated by a telescope, the pair is called a *visual binary*. We humans know close to 100,000 visual binaries, and that star we are looking at now, Mizar, was the first visual binary star discovered. In 1650, the Italian astronomer Giovanni Battista Riccioli studied Mizar and could distinguish two components revolving around each other. Later, humans looked at stellar spectra and realized that Mizar is actually a quadruple star! In the case of Mizar, each of the optical components is a *spectroscopic binary star*. We can't resolve these kinds of pairs with a telescope, but we can by observing a varying Doppler shift in the spectral lines as the stars approach and recede from Earth. We know of over a thousand spectroscopic binaries."

Bob points to the ground, "Linux pods."

"Linux what?" says Miss Muxdröözol.

"The plants, look down." Fragile trailing vines with small pink and pale blue flowers taper to form pods, each with a number of seeds.

"Sir," Mr. Plex says, "can we eat them? I'm starved."

"No," Bob says, "but underground, below the flower at the bottom of each stem is a seed we can eat. I'm familiar with this species."

Bob pulls a handful up from the ground, wipes the dirt off on the violet grass, and pops seeds into his mouth. "A little gritty," he says, "but not bad."

"Hey," Miss Muxdröözol says, "you should be careful about what you eat."

"Remember, I studied botany in school."

Miss Muxdröözol stoops lower. "Yeah, but is this plant really safe?"

"Try some."

Miss Muxdröözol bends down and pulls up several seeds. "I'll starve if I just stand here watching you." She pauses and makes a strange face. "They taste dry."

Mr. Plex makes a sick crunching sound as he pops dozens of nuts into his mouth simultaneously. "Yes, let us hope we find something better to eat."

"I've told you about visual binaries and spectroscopic binaries. An *astrometric binary* is a coupling with a partner star totally invisible to us."

Miss Muxdröözol continues to chew on a nut. "If the star is totally invisible, how do you know it is there?"

"We can guess that an invisible partner exists by observing the periodic back and forth motion of the visible star. The bright star Sirius in Canis Major was first thought to be an astrometric binary until 1862, when telescopes helped us to see that Sirius was really two stars, Sirius A and a faint Sirius B. We already talked about how Sirius is the brightest star in the night sky, with apparent visual magnitude −1.5. The bright partner, Sirius A, is a blue-white star 23 times as luminous as the Sun. It's 8.6 light-years away from Earth, only about twice the distance away as the closest star, Alpha Centauri. This dim partner, Sirius B is about as massive as the Sun, though much more condensed. It was also the first white dwarf star to be discovered." (See figure 5.5.)

"Sir, I first got interested in stars in the late 1960s by watching the TV series *Lost in Space*. The show describes the travels of a human family exploring strange planets. Their mission was to begin colonization of a planet near the star Alpha Centauri. Unfortunately, their craft went off course, and they lost all contact with Earth."

"Mr. Plex, you were monitoring our TV broadcast from afar? Well let me tell you something that the TV show never revealed. Alpha Centauri is actually a triple star, the faintest component of which, Proxima Centauri, is the closest star to the Sun, at about 4.3 light-years distance. The two brighter partners are one tenth of a light-year farther from the Sun. They revolve around each other within a period of 80 years, while Proxima orbits them within a period probably of millions of years. Gets kind of complicated, doesn't it? Danger, Will Robinson."

"Who is Will Robinson?" says Miss Muxdröözol.

"Never mind. Let me finish up and tell you about *eclipsing binary stars*. Sometimes we detect binary stars by changes in apparent brightness, as the darker (or dimmer) star blocks its brighter partner at regular intervals. For example, the star Algol in Perseus seems to wink at you every two days because one star cuts off light from the other as the pair orbit about one another. Finally, there are stars called *optical doubles*. But they are really just frauds."

"Frauds?" says Mr. Plex.

"Yes, they're a pair of stars that *appear* to be close to each other when you look up at the sky, but in reality they're far away from each other. There's no gravity holding them to one another like in actual double stars."

"What's the oddest coupling you know of?" asks Miss Muxdröözol.

Bob raises an eyebrow. "My favorite is Castor, in the Gemini constellation. It has at least six component stars in a crazy gravitational orgy. Though Castor appears as a single star to the naked eye, a telescope resolves this into two closely spaced stars, Castor A and Castor B, and a distant, dim red companion named Castor C."

Bob places three stones on the ground to represent Castor A, B, and C:

    ✷ ✷   ✷

Castor A and B   Castor C

"Gravity binds A, B, and C together so that they orbit around each other. Castor A and B are separated by about the same distance as the diameter of our solar system, and they revolve around each other every 500 years. The lonely Castor C completes its orbital journey in ten millennia."

"Sir, you said Castor has six components," says Mr. Plex.

"Yes, the Castor star system has many surprises for us. We can't see any more stars using a telescope, but spectroscopic analysis of the light coming from Castor A, B, and C reveals that each of the three visible stars in Castor is itself made up of a pair of nearly identical stars." Bob places two stones on the ground to represent Castor A.

✹ ✹

Castor A

"The two stars of Castor A are only four million miles apart, about sixteen times the distance between the Earth and the Moon. That's amazing. Try to imagine the Castor binary with each star about twice the size of our Sun, close together, and spinning madly around each other once every nine days! The pair of stars that make up Castor B are only three million miles apart and orbit each other in just three days. And finally, the two eclipsing red dwarf stars of Castor C are only six Earth-Moon distances apart and spin around their common center of gravity in only 19 hours! They orbit the four main stars in a period of millions of years. This is the ultimate cosmic juggling act."

"I get dizzy just thinking about it," says Miss Muxdröözol.

"What would Galileo have thought of Castor? What would the older Christian Church have made of such an odd coupling given their simplistic heliocentric (Sun-centered) Universe?"

"Good question," says Mr. Plex.

"Listen," Miss Muxdröözol says. "Voices are coming from behind those trees. Sir, there are some people there, I think. Maybe they can help us get off this asteroid."

"But to where? We're in the future. Everyone we knew and loved is long gone."

"Sir, we can't stay here forever. We have to find out more about this place," says Miss Muxdröözol.

"Wait, get down," Bob says, hiding himself within the long, shimmering blades of grass. "We don't know who or what they are. Let's first get a closer look before we introduce ourselves. Okay? Follow me."

As Bob crawls through the damp violet grass, getting closer to the source of the voices, he hears squishy sounds in the dirt. He passes large holes in the ground, about five feet in diameter. A many-fingered swirl of smoke puffs up from one of them. The smoke is lime green, like the color of the iceworm paintings in his room at the art museum—a museum long gone.

"Look!" Miss Muxdröözol says.

Bob sees a horrified expression on Miss Muxdröözol's face. "Look at them," she says. Her whole body tightens, and then she takes a breath. "They're—they're like giant ants."

Bob looks beneath the dogwoods at the two creatures. Each has two large multifaceted eyes. They never blink. Their six limbs are jointed in three places, and their stride lengths are close to five feet. One of the ants has a golden abdomen. The other has an abdomen that blinks red and green, like a Christmas tree light.

"Did you hear about the visitors?" says the golden one using whatever alien vocal apparatus is jammed into its elongated head. Its voice is a whispering of sibilants.

Before the other ant responds, the gold one removes an electrical probe from a pouch and twists it up a single hideous hole in the center of its face. Bob put his fingers over his nose. The creature's body exudes the ammoniac smell of piss and sweat.

"Yes," replies the blinking ant, "there should be three of them, a man, a scolex, and a female of unknown origin." His raspy voice is curiously similar to the larger creature's—a regional accent of a language other than English. Each has two large jaws protruding from the heads. Some of their limbs terminate in claws. Others seem to end with fingers, all the same length, with no obvious thumb. Two of the legs end in pads of iridescent hairs. Their eggplant-sized heads sit on large saclike abdomens.

"What do we do with them, when we find them?" hisses the blinking ant with an androgynous voice.

"You mean *if* we find them."

The blinker waves one of its slender, sinewy legs. "Right, *if* we find them."

"Bring them to the Nephilim. They'll tell us what to do."

One of the ant things shambles over to a dogwood tree and clasps a branch with its claw-like hands. It shakes a large beetle from the branch into its mouth. "Try one," it says.

Bob whispers to Miss Muxdröözol who lies beside him in the waving field of grass. "This is impossible!" Bob says. "They're bloody *ants* for cryin' out loud! They shouldn't even have vocal cords." Then he calms down as a hard fist of fear grows in his stomach. "Let's quietly get out of here." Bob's pants are now thoroughly soaked with mud. He shifts his position to a drier patch of ground.

"Wh— what *are* those things? Miss Muxdröözol says, as she begins to retreat.

"Shh!" says Bob. "Wait, we've got to hear what they say. We need more information about this place if we want to survive or find a more suitable world or place to live."

The creatures start shuffling down a path, away from Bob, Miss Muxdröözol, and Mr. Plex. Soon Bob cannot see them.

Some Science Behind the Science Fiction

Time is but the stream I go a-fishing in. I drink at it; but while I drink, I see the sandy bottom and detect how shallow it is. Its thin current slides away, but eternity remains. I would drink deeper; fish fill the sky, whose bottom is pebbly with stars. I cannot count one. I know not the first letter of the alphabet. I have always been regretting that I was not as wise as the day I was born.

— Henry David Thoreau (1817–1862), *Walden*

In this chapter, Mr. Plex and Bob discussed using cosmic microwave background radiation to estimate the age of their Universe. They also used the radiation to estimate how far they traveled into the future. This radiation is the long wavelength electromagnetic radiation left over from the Big Bang, the explosion at the beginning of the Universe. The microwave part of the electromagnetic spectrum is equivalent to the shortest wavelength radio waves. To better understand microwave background radiation, let's first consider how light behaves. Recall from chapter 2 and the "electromagnetic piano" figure that light and microwaves are just different parts of the electromagnetic spectrum (figures 2.4 and 2.5).

Light travels at 299,792.45 km/sec (186,282.396 mi/sec). When you look at stars in the sky you are looking into the past, because the light hitting your eye today left the star years ago. Most of the stars you can see are 10 to 100 light-years away. This means you are seeing the stars as they looked 10 to 100 years ago.

Imagine you are looking up at the Andromeda Galaxy, the nearest giant Galaxy (aside from our Milky Way) to Earth. This beautiful spiral galaxy is one of the few galaxies you can see with your naked eye. (However, without a telescope it only appears as a milky blur.) Because the Andromeda galaxy is located about 2 million light-years from the Earth, when you look at it, you are seeing it as it was 2 million years ago. For all we know, some giant monster could have already swallowed Andromeda, but we wouldn't know until 2 million years later. Astronomers observing distant galaxies with the Hubble Space Telescope can see them as they were only a few billion years after the Big Bang.

By observing the cosmic microwave background, cosmologists are looking far back in time and gleaning information about the Universe only a few hundred thousand years after the Big Bang, long before stars or galaxies even existed. By studying the background radiation, we can learn about the structure of the Universe, its origin and evolution.

One of the fundamental predictions of the Big Bang theory is that the Universe is expanding. Astronomers who examine distant galaxies can directly observe this expansion, originally detected by Edwin Hubble. Since the Universe is expanding, it is colder and less dense than it was in the distant past. In fact, when the visible Universe was half its present size, the density of matter was eight times higher, and the cosmic microwave background was twice as hot. When the visible Universe was one hundredth its present size, the cosmic microwave background was a hundred times hotter, which made it about the temperature at which water freezes. When the visible Universe was one hundred millionth its present size, its temperature was 273 million degrees above absolute zero, and the density of matter was comparable to the density of air at the Earth's surface. At these high temperatures, the hydrogen was completely ionized into free protons and electrons. Not until about 500,000 years after the Big Bang did the Universe cool sufficiently to permit protons and electrons to combine to form neutral hydrogen.[6] The Big Bang created helium and the light elements in the first few minutes of the Universe's existence, providing some raw material for the first generation of stars. However, stars were needed to create the heavy elements and spit them out into space when a star exploded or when a red giant sheds some of its outer atmosphere.

Sirius, the brightest star in the night sky, has been mentioned several times in this book. The star, located in the constellation Canis Major, is also known as the Dog Star (see figure 5.5). According to ancient mythology, this constellation represented a dog following the heels of the Greek hunter Orion. In 1844, the German astronomer Friedrich Wilhelm Bessel observed that the star had a slightly wavy course of travel and concluded that Sirius was actually a binary star. In 1862, Alvan Clark, an American astronomer and telescope maker, was the first to see both of these stars.

The ancients were well aware of this Sirius. The ancient Egyptians called the star Sothis, and they believed that Sothis caused the Nile floods after observing that the star made its first rising of the year at the time of the Nile's annual floods. Because the Egyptians considered Sirius the herald or cause of the Nile rising and of a subsequent good harvest, they built their temples in such a way that the light of Sirius reached the inner chambers. The Egyptians represented the star as a five-pointed star at the famous Temple of Isis at Denderah, and all the ancient civilizations—the Mesopotamians, Akkadians, Babylonians, Assyrians, Chaldeans, and Persians—knew of its existence.

The hottest part of the summer in the Northern Hemisphere coincides with the "heliacal" rising of Sirius. That's where we get the phrase "the dog days of summer." The heliacal rising of a star occurs when, after being in conjunction with the Sun and invisible, the star emerges from the light so as to be visible in the morning before sunrise.

* * *

Stars on the upper left of the Main Sequence are massive dwarfs. Stars on the lower right of the Main Sequence are light dwarfs (see figures 5.1 and 5.3). Recall that at the end of chapter 4, we discussed how stellar lifetime is inversely proportional to the mass of a star or more specifically, $\tau \propto (1/M)^{2.5}$. A rare O-class dwarf at the upper left of figure 5.1 survives for three or four million years, which means that some stars have come and gone since humankind walked the Earth.[7] On the other hand, a cool M dwarf at the bottom right of the Main Sequence can live for trillions of years. Stars on the lower main sequence have never had time to die. Every one that ever was is still here today.[8]

A magnification of the H-R diagram in figure 5.3 can be quite instructive. Figure 5.8 shows the 50 nearest known stars to Earth, with the sizes of the stars indicated by the size of the circles. Looking at the plot, you can see that most of the Sun's nearest neighbors are relatively small stars, the slow-burning red dwarfs.

* * *

The star Betelgeuse, described in this chapter, is indeed a big one, 54,000 times brighter than the Sun and large enough to extend to the orbit of Mars if placed at the center of our Solar System. But because the big guy has a lower surface temperature than the Sun, it produces a lower solar flux per unit of surface area. However, Betelgeuse is not the limit in star size. For example, consider the star Mu Cephei in the constellation Cepheus. Mu Cephei is 1.7 times larger and has a radius comparable to the orbit of Saturn. You could placed over one billion Suns inside Mu Cephei. (Megastars are discussed further in appendix 2.)

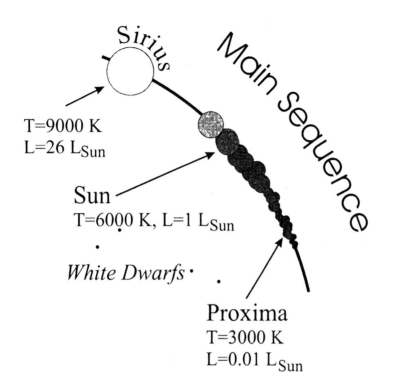

T=9000 K
L=26 L$_{Sun}$

Sun
T=6000 K, L=1 L$_{Sun}$

White Dwarfs

Proxima
T=3000 K
L=0.01 L$_{Sun}$

Figure 5.8   The 50 nearest known stars to Earth, displayed schematically on a
magnified H-R diagram (see figure 5.1). Star size is indicated by the size of the circles.
Luminosity is indicated by the gray shades. Luminosity also increases along the vertical axis.
Temperature increases from right to left. Stars on the Main Sequence have the proper
configurations to support fusing hydrogen into helium. Massive stars dominate the light
output of the Galaxy. [After Fred Adams and Greg Laughlin, *The Five Ages of the Universe*,
(New York: Free Press, 1999), 43.]

"My greatest nemesis still provides our customers with free light, heat
and energy. I call this enemy . . . the Sun. Since the beginning of time,
man has yearned to destroy the Sun, I will do the next best thing . . .
block it out.

— Mr. Burns, in *The Simpsons* TV show, "Who Shot Mr. Burns? Part 1"

☆ Chapter 6

Last Tango on the Heliopause

When to the new eyes of thee
All things by immortal power,
Near or far,
Hiddenly
To each other linked are,
That thou canst not stir a flower
Without troubling a star.

— Francis Thompson, *The Mistress of Vision*

"Sir, Miss Muxdröözol's legs are gone. Someone must have stolen them."

Bob reaches down to help Miss Muxdröözol out of a hole she has fallen into while hiding from the alien ants. "Mr. Plex, she still has her legs. It's normal for her species to lose legs and regenerate them minutes later. She has body-repair nanoparasites living in the left acetabulum of her pelvis, or so I'm told. Give her a minute, and she'll be good as new."

Miss Muxdröözol nods her head and smiles, her bioluminescent canines sparkling against a stellar backdrop. Her legs have started to regenerate.

Bob feels a warm breeze become stronger. It is the kind of wind that roars in their ears and blows dirt into their hair and between their teeth. Then the breeze subsides and the air is suddenly still. The ants have disappeared, at least for the moment.

"Mr. Plex," Bob says, "do you have any idea who they are?"

"The ants?" says Mr. Plex.

"Of course, the ants."

"They seem to be after us. Maybe they don't want us here. But we have nothing to worry about given your superior fighting skills and your ten years of training in Kung Fu."

"True. Therefore, I suggest we finish up our lesson on stars so that we can prepare to journey forward to the end of the Universe. That membrane creature took us millions of years into the future. If it comes back, it may take us even further into the future. That's my goal now. I want to see what happens in trillions of years from now. I want you both to see for yourselves how stars evolve and die, like silent embers fading in a funeral pyre."

"Very poetic, Sir."

"There seems to be ample water here for us to survive, provided the grass is edible and we find other food sources." He runs his hands through the grass, "Not a bad place to die." Bob motions Mr. Plex and Miss Muxdröözol to sit besides him in the grass.

"Today I want to talk about the Sun. If you understand how the Sun works, you'll be a long way toward understanding stars in general. If you recall, when we discussed stellar luminosity and distance modulus, I told you that the Sun's luminosity was 100 septillion watts. A lot of the energy is lost in space. The amount of energy that hits the Earth's atmosphere per second per unit area is about 1,370 watts per square meter or 126 watts per square foot. The value can change very slightly with variations in solar activity, but any big change would wreak havoc for life on Earth."

Miss Muxdröözol sits closer to Bob. "That still sounds like a lot of energy hitting the atmosphere."

"Yes, but only a percent or two is captured by plants using photosynthesis. That's because almost 99 percent of the solar energy reaching the Earth is *reflected* from leaves and other surfaces and absorbed by other molecules, which convert the light to heat. About 50 percent of the sunlight that reaches the ground is visible light. The rest is mostly infrared radiation and smaller amounts of ultraviolet light."

"All that energy traveling through space," Miss Muxdröözol says. "And the Sun is about 100 million miles away from Earth?"

"More precisely, the Sun is about 150 million kilometers (93 million miles) away from Earth, on average. Astronomers call this distance an *astronomical unit* or *AU*. As a comparison, the average distance between Jupiter and the Sun is about 5.2 AU, and the average distance between Pluto and the Sun is about 40 AU."

Mr. Plex begins to chew on the grass. Violet juice runs from his mouth. "Sir, not bad. Tastes a bit like what your people call poomgrans."

"Pomegranates?"

"Yes, pomegranates," Mr. Plex pauses. "Sir, how did the solar system and planets form?"

Bob places a single blade of grass in his mouth and begins to chew. It does have a fruity taste. He hopes it is safe. He knows that certain alien species of grass are capable of producing some of the most toxic substances known. A species of grass found in deep subterranean chambers on Ganymede produced in humans a burning sensation, followed by a series of ecstatic religious visions often involving angels, Moses, and unleavened bread, and eventually caused death.

Bob pushes the thoughts out of his mind and enjoys the grass. "For centuries, scientists hypothesized that the Sun and planets were born from a rotating disk of cosmic gas and dust. The flat disk constrained the nascent planets to have orbits lying roughly in the same plane. This *nebular theory* was proposed as far back as 1755 by the philosopher Immanuel Kant.

Bob puts another blade of grass in his mouth. They're hard to resist.

"Kant was essentially correct," Bob says. "In the last years of the twentieth century, the Hubble Space Telescope (HST) revealed several dozen disks at visible wavelengths in the Orion Nebula, a giant stellar nursery about 1,600 light years away. We call them 'proplyds,' a contraction of the term protoplanetary disks. The Orion proplyds are larger than the Sun's solar system and contain enough gas and dust to provide the raw material for future planetary systems"[1] (figure 6.1).

Now Miss Muxdröözol is chewing on the grass. "So Kant got it mostly right in the 1770s?"

"Brunhilde, display solar nebula theory." Brunhilde displays figure 6.2. "Yes, stars and their disks form by the gravitational collapse of large volumes of sparse

Figure 6.1   Attack of the proplyds! A Hubble Space Telescope close-up of the Orion Nebula reveals disks of dust and gas surrounding newly formed stars. These fuzzy blobs, called "proplyds," may be infant solar systems in the process of formation. This chunk of the sky spans about 0.14 light-years. (Image courtesy of Charles Robert O'Dell of Vanderbilt University and NASA.)

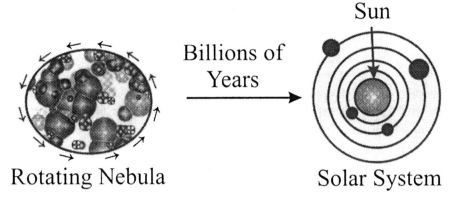

Figure 6.2   Solar nebula theory posits a rotating nebula condensing to form a Solar System.

interstellar gas called solar nebulae. Sometimes a shock wave from a nearby supernova, or exploding star, may trigger the collapse."

Bob runs his fingers through the violet grass. It feels soft, like satin. "Why do you think all this grass is here? There's so much of it."

Miss Muxdröözol swallows some of the grass in her mouth. "Maybe it's here to suck up carbon dioxide or some other gaseous waste."

"Sir, maybe it's a food source for the ants," says Mr. Plex.

Miss Muxdröözol nods. "I have an idea. This place is obviously artificial, created for us, other travelers, or the ants. Maybe they like the way it feels while walking through it—a sensory delight. We don't know much about those ants, but if they are trapped on this asteroid for many years, maybe they grew the stuff to boost their spirits." Miss Muxdröözol begins to lie down and crawl through the grass as it slaps up against her face. "Ah, I like the way the grass feels."

Bob looks into Miss Muxdröözol's dilated eyes. "Let's talk about the Sun. Here are the facts. When we were back on Earth, astronomers had identified more than 60 elements in the Sun's spectrum. The Sun's *outer* layers had the same ratio of elements that the Sun had when it formed: 72 percent hydrogen, 26 percent helium, and 2 percent other elements by weight. However, in the twenty-first century, the Sun's core had around 40 percent helium because some of the hydrogen was converted to helium in nuclear fusion reactions."

Bob reaches into his pocket and tosses an orange to Mr. Plex. "I want to talk about the Sun's structure. The Sun has an *atmosphere* consisting of three layers: the *photosphere, chromosphere,* and *corona.* The photosphere is the visible surface of the Sun. It's about 400 kilometers (250 miles) thick. That's a bit more than the distance from New York City to Washington D.C."

"Sir, where did you get the orange?"

"Hieronymus always insisted I carry a small orange for emergencies." Bob pauses. "Mr. Plex, I want you to think of the orange rind in your claw as the photosphere. Most of the light that reaches Earth comes from the photosphere because light generated deeper inside the Sun can't escape without absorption and reemission. I'll explain what that means in a minute." Bob pauses and says, "Brunhilde, display photosphere." Brunhilde displays figure 6.3. "The photosphere is pretty tenuous stuff—the density is about one thousandth that of the air we breathe. Sunspots—dark blotches on the sun—occur on the photosphere. To visualize a sunspot, imagine a dark moldy spot on the orange skin. Telescopic observations reveal that the photosphere has a grainy appearance, like Styrofoam. Brunhilde, please display solar structure table."

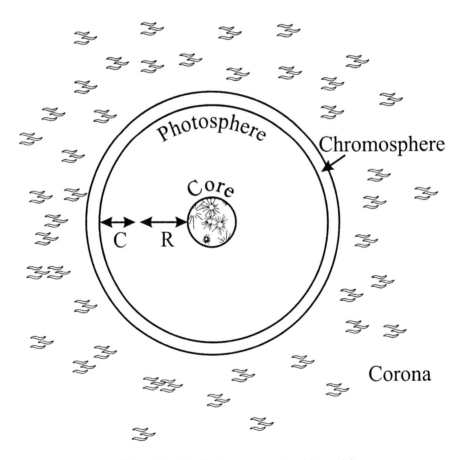

Figure 6.3   The Sun's structure. C and R stand for
convective zone and radiative zone, respectively.

| | Table 6.1 | |
| | The Sun's Structures | |
Region	Temperature	Layer Thickness
Interior ("core")	15 million degrees K	175,000 km
Photosphere	5,800 degrees K	400 km (250 miles)
Chromosphere	15,000 degrees K	10,000 km (6,000 miles)
Corona	2 million degrees K	Millions of km into space

Bob begins to walk and notices a few small rodents with weblike feet. At least there will be meat to eat if he has to survive here for while. "When astrophysicists examine stars like the Sun, they observe that the brightness decreases away from the center of the Sun to its apparent edge, called the *limb*. This limb darkening occurs because the photosphere's upper, cooler layers produce less light than the center. Right above the photosphere is a layer called the *chromosphere*. It contains a lot of hydrogen and is surprisingly much hotter than the photosphere. The *corona* is above the chromosphere. It extends millions of kilometers into space and shines at X-ray wavelengths."

Mr. Plex returns the orange to Bob who takes a pocket knife from his pocket and begins to cut away at the orange. He tosses the skin of the photosphere to Mr. Plex, who promptly begins to chew on the rind.

"Mr. Plex, you are chewing on the photosphere. It is not edible."

"Edible?"

"Never mind. In the center of the Sun is the core, the Sun's power plant in which nuclear fusion reactions turn hydrogen into helium and generate tremendous amounts of heat. Here, the gas density is more than 100 times that of water, or 14 times that of lead. In fact, the core contains 40 percent of the solar mass."[2]

"Sir, at that density, why isn't the core a solid?"

"The very high temperatures keep the core a gas. At the center, about half of the hydrogen had already been fused into helium in the twenty-first century. All the energy released in the core exerts an outward pressure against the inward pull of gravity. (It's the pressure gradient that produces the force to oppose gravity.) Lots of photons are produced in the Sun's interior, but they are absorbed and reemitted at lower energies in the surrounding *radiation zone*. This was marked as "R" on the previous figure displayed by Brunhilde. The energy in the core escapes the Sun by a random-walk process, like a drunk who walks from Canada to Mexico by taking random roads along the way."

As Bob and his friends travel, they see a few more of the rodents, but nothing else. There are no birds, large trees, or hills. "It takes about 30,000 years for the energy in the core to finally leave the Sun. At the *convection zone*, beneath the photosphere, the Sun's gases begin to roll in a complex series of layers that eventually come to the surface and are seen as *granulation*. This is the Styrofoam appearance I mentioned, or perhaps a more apt analogy is the mottled complexion of the orange rind dangling from your mouth. You can see the solar granules using the proper kinds of telescopes. Don't ever observe the sun without taking proper precautions, because you could go blind. The convection zone transfers the energy as heat from lower layers to outer layers. Brunhilde, display the surface granules on the sun followed by the surface of an orange." Brunhilde displays figures 6.4 through 6.6.

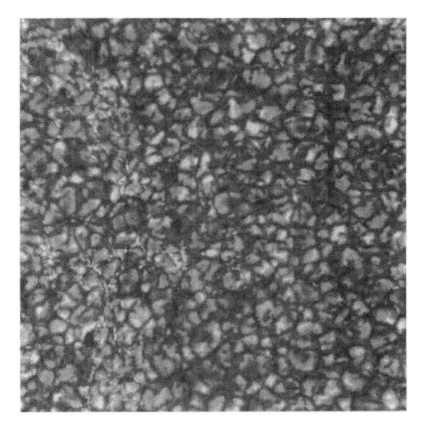

Figure 6.4   High-resolution photograph of solar granulation with granules in different stages of evolution. Several granules have dark, cool centers, a precursor to their disintegration, while others can be seen in the process of splitting. The smallest bright features between granules are connected with magnetic structures. This image was recorded by Göran Scharmer (Royal Swedish Academy of Sciences, Stockholm Observatory) using the Swedish Vacuum Solar Telescope on La Palma in the Canary Islands, shown in figure 6.5. (Figure courtesy of Göran Scharmer.)

Figure 6.5   The Swedish Vacuum Solar Telescope on La Palma in the Canary Islands.
This device was used to capture the granulation image in figure 6.4.
(Figure courtesy of Göran Scharmer.)

Figure 6.6   Surface of an
osage-orange, reminiscent of
the granulated surface of the
sun (figure 6.4). [From Harlow,
Art Forms from Plant Life (New
York: Dover, 1976), 69.]

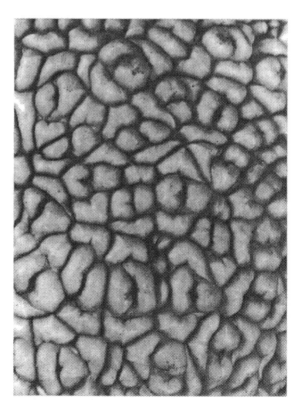

"What is *that*?" Mr. Plex says as an insect rises from the grass. It drifts towards Miss Muxdröözol then to Mr. Plex, moving like a feather floating in the breeze. Mr. Plex steps back. His eyelids peel back in wonder. The large dragonfly, floating at the height of his head, flashes various shades of violet and crimson. At the edges of the membranous wings are little sparkles, as if the light is igniting tiny dust particles in the air as the wings move through them.

Miss Muxdröözol comes closer to Bob. "You said that it takes over a million years for the energy in the core to leave the Sun? How could that be."

"I told you about the Sun's radiative zone that surrounds the core. This zone is dense with ions and energy, but there isn't enough heat or pressure to cause hydrogen to fuse. Instead, the matter in the radiative zone just gets in the way of energy from the core. Each photon, or packet of light, produced in the core moves into the radiative zone where it is absorbed, reemitted, deflected, and bounced around like a ball in a pinball machine. The random movement of each photon through the radiative zone delays the energy from escaping."

"I find solar granules[3] among the most fascinating objects in the Universe. Here are some awesome facts." Bob pulls out a piece of paper from his pocket and hands it to Miss Muxdröözol.

Solar Granule Fact Sheet[4]

* A solar granule is about the size of the state of California.
* A solar granule (bubble) is 400 degrees K warmer than the dark intergranular network of lanes that surrounds the granule.
* The lifetime of an average granule is ten minutes.
* Solar granules are like bubbles on the top of boiling tomato soup. Heat is supplied from below and carried upward through the process of convection. The cool liquid on top of the soup sinks.
* About 40 percent of the solar granules explode, vomiting their material into the solar atmosphere at speeds of 1,000 miles per hour (or about 500 meters per second). If we could hear sound waves across the vacuum of space that separates us from the Sun, we would hear a thunderous roar accompanying each granulation eruption.
* Many granules together form *supergranules*, that is, large convection cells around 19,000 miles (30,000 km) across, much larger than the Earth's diameter. Supergranules last several hours. Jets of gas called *spicules* erupt upward at the edges of the supergranules. They can be as long as the diameter of the Earth. *Faculae* are bright, white hydrogen clouds that form above regions where sunspots are about to form.[5]

"Oh my God!" Mr. Plex says as he points to another pair of iridescent wings rising from the murky swamp. The creature seems to follow the first insect.

Bob ignores him. "The Sun also has sunspots, which are dark, irregularly shaped regions on the Sun's photosphere. They are actually solar magnetic storms."

"Why are the spots darker?" Miss Muxdröözol smiles as the twin insects dance in front of her eyes. She whistles and doesn't seem frightened at all.

"Because, the temperature of the spots is lower than that of the surrounding photosphere. However, they're still very bright, brighter than many cool stars. And although they are cooler than the rest of the photosphere, they are still so hot that they would melt any substance on Earth. Most sunspots have a dark central portion called the *umbra* and a lighter outer area called the *penumbra*. It reminds me of an eye, in which the dark pupil is the umbra and the iris is the penumbra. But it's not like any human eye—the pupil, or umbra, would be 4,200 degrees K and the iris would be several hundred degrees cooler than the photosphere." Brunhilde, of her own volition, displays an ☞ followed by a close-up image of a sunspot (figure 6.7).

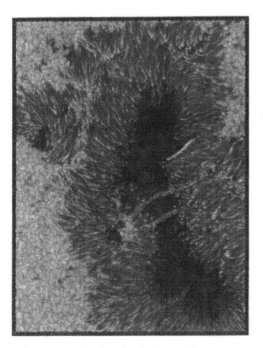

Figure 6.7   Close-up of a sunspot showing the dark central umbra, surrounding penumbra, and granulation of the photosphere. The image was acquired using the adaptive optics system at the Swedish Vacuum Solar Telescope (figure 6.5) Other features may be seen, including umbral dot, two light bridges in the dark umbra, bright grains, dark clouds, and bright and dark filaments in the penumbra, pores, and disconnected penumbrae.
Tick marks are 1 second of arc (approximately 700 km on the Sun).
(Image courtesy of Göran Scharmer and Luc Rouppe van der Voort.)

Bob tosses a magnet to Mr. Plex, "Shove that magnet beneath the rind of the orange in your claw."

"Yes, Sir."

"That's a good representation of a sunspot. Even though people can see sunspots in the visible spectrum, sunspots are also magnetic regions with field strengths thousands of times stronger than the Earth's magnetic field. And you'd need a bar magnet as big as Earth to simulate a sunspot. Many sunspots come in pairs with a magnetic field that looks somewhat like that of a bar magnet. Magnetic lines of force escape from one spot (the North Pole), loop overhead, and dive back down to the other member of the pair (the South Pole). The field is strongest in the dark umbra" (figure 6.8).

"Be careful," Bob says to Miss Muxdröözol. "Move away from those dragonflies."

Miss Muxdröözol moves her hand. "Didn't your mother ever tell you that dragonflies can't bite? Their mouths are *tiny!*" For a moment Miss Muxdröözol reminds Bob of his first girlfriend. He was at college when he first passed her in

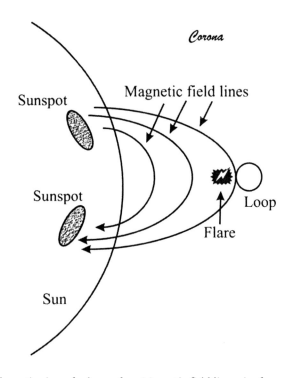

Figure 6.8   Schematic view of solar surface. Magnetic field lines rise from one sunspot into the corona and plunge back down into the other. At the top of the loop, the filed lines can become twisted to produce a solar flare that lights up the chromosphere below.

the hallway. She was wearing metallic bell-bottoms and a shirt full of light-emitting diodes, and she winked at him. At seven-second intervals, the lights spelled out her name: Neuron, Neuron, Neuron. Ah, but now Neuron's beautiful bones were long turned to dust. But, at the time, Bob was an idealist, always trying to make the Earth a better place, but not for himself. It was always for her, Neuron, and the future of humanity.

Bob shrugs. "A huge release of energy—a solar flare—can occur if the magnetic field is twisted above the sunspot. Solar flares are the biggest explosions in the solar system. They can release the same amount of energy as millions of 100-megaton hydrogen bombs all exploding at the same time."[6]

"Where is all the energy coming from?" Miss Muxdröözol asks.

"The energy comes from the Sun's magnetic field. Think of the twisted magnetic field around a sunspot group as a huge stretched rubber band that's ready to fly. And when the solar flare goes off, it can affect Earth, overloading electrical power grids and destroying satellites when their electronics are bombarded by charged particles. Way back in 1989, a solar flare induced electrical currents on the ground that caused the Hydro-Quebec electric power system to fail."[7]

Mr. Plex grinds the magnet into the orange rind until Bob can see it make a dent. "Mr. Plex, that's also a good metaphor for sunspots. They can be magnetically detected before you can see them, and after they visually disappear from the photosphere, the magnetic field persists. In addition to these sunspots' local magnetic fields, the entire Sun also has a magnetic field."

Bob pauses to watch the dragonflies. "A *magnetograph* is an instrument that allows us to observe the Sun's magnetic field. Harold and Horace Babcock, father and son, invented the device in 1951, and they also discovered magnetically variable stars. In the late 1950s, Harold Babcock confirmed that the Sun reverses its magnetic polarity periodically."

"Sir, do sunspots exist on other stars?" asks Mr. Plex.

"Starspots similar to sunspots seem to exist on other stars, as indicated by brightness variations and spectral line measurements."

Bob takes the magnet from Mr. Plex's claws. "The Sun's magnetic field extends throughout the solar system. It's quite complicated because of the Sun's rotation and the convection of electrically charged particles in the Sun's hot gases. The magnetic field reverses every 11 years, right after the time when the sunspot number peaks."

One of the dragonflies continues to dance before Miss Muxdröözol's eyes. The other slips between her legs, rises along her back, and nearly collides with the first insect. Miss Muxdröözol's grin widens, and her whole body seems to shiver like an excited child's.

"The magnetic activity that accompanies the sunspots seems to produce dramatic changes in the ultraviolet and X-ray emission levels. These changes over the solar cycle can affect the Earth's upper atmosphere.[8] Changes in the Earth's magnetic field are known to be caused by solar storms, but the precise connections between them and solar activity levels are still uncertain.[9] Planning for satellite orbits and space missions often requires knowledge of solar activity levels years in advance."[10]

"Sir, you mentioned that sunspot activity reaches a maximum once every 11 years. What does that mean?"

"Sometimes the Sun has hundreds of sunspots. At other times, it has no sunspots. The number of sunspots regularly reaches a maximum and minimum every 11 years. This cycle is called the *sunspot cycle*."

A few strands of Miss Muxdröözol's hair stand at right angles to her body; perhaps there is static electricity in the air. She raises her hands as if she is conducting an invisible orchestra, but to Bob it begins to look more like she is casting a spell or stirring a witch's potion.

"The Chinese knew about sunspots before 800 B.C. When Galileo used a telescope to study sunspots, some religious authorities of his day could not believe his report."

"Why not?" asks Miss Muxdröözol.

"They thought the Sun had to be perfect and that spots would be an insult to God. Incidentally, there is evidence that solar activity can affect our weather and climate. For example, from 1645 to 1715, there was a period (called the Maunder minimum) when there were hardly any sunspots, and Earth's weather was unusually cold."

"Not a coincidence?" Miss Muxdröözol says.

"Not a coincidence. It was a period known as the Little Ice Age. Similarly, the Sun appears to be most active when the sunspots are most numerous. Astronomers try to understand the sunspot cycle in order to make more reliable weather forecasts."

Mr. Plex stares at his orange rind. "How big are sunspots?"

"Each sunspot is huge, about the size of Earth and larger. They can last for days or months as they are carried around by the rotation of the Sun."

Miss Muxdröözol rolls in the grass. "Please excuse me for a moment," she says.

"What's she doing?" Bob says to Mr. Plex.

"Shedding her skin."

"Like a snake?" Bob gasps.

In minutes, Miss Muxdröözol emerges as white as a pearl. Her hair looks like ice.

"Du bist wie eine Blume!" Bob whispers to his antediluvian gods. *She is a surprise a second.*

"Please continue," she says. "Don't let the peculiarities of my deliquescent skin disturb you."

"'Deliquescent'—that's a fine-sounding word," Mr. Plex says.

Bob lets out a great gust of air. "Let me tell you about the physical laws operating on the Sun that let life evolve. Generally speaking, the Sun's interior fuses four atoms of hydrogen into one atom of helium. The number of atoms decreases in this process, and this means that the pressure of gas decreases because pressure depends on the number of atoms in a given volume.

"Sir, if the pressure in the center decreases, doesn't that make the Sun's core contract under the weight of the rest of the Sun?"

"Yes, but as a result of the contraction, the temperature of the core increases, so the fusion rate and luminosity continue even though the fuel supply decreases. It is this process of contraction and temperature increase that keeps the Sun stable over billions of years. Without it, life wouldn't have had time to develop, and you, Mr. Plex, would not be here talking to me."

Bob tries not to stare at Miss Muxdröözol's pearl skin. "The Sun's Main Sequence lifetime is about 10 billion years. When the fuel is used up and hydrogen is no longer being fused, the Sun will blossom like a great red rose. The resulting red giant will engulf the Earth, but let's talk about that later. That won't happen for another 5 billion years. As my friend the astronomer James Kaler once wrote, 'Nature has produced a wonderfully balanced fusion machine, with the unimaginable violence in the core so tightly controlled that we will be able to go outside and enjoy a sunny day for a long time to come.'"[11]

Miss Muxdröözol looks up at the sky. "Ah, I wonder if we will ever get off this asteroid."

Mr. Plex tosses the orange to Miss Muxdröözol, and the orange spins as it travels. "Sir, the Sun spins on its axis from west to east, just like the Earth."

"That very learned of you, Mr. Plex. However, there's an important difference. Every day the entire Earth makes a complete rotation, but that's not true with the Sun. The Sun spins fastest at its equator and makes a complete rotation at the equator in about 25 days. It spins slower at the poles so a compete rotation there takes 35 days." Bob tosses the orange to Mr. Plex.

"Sir, how is that possible? There's no way I can spin different parts of this orange at different rates."

"C'mon. The answer is obvious. The orange in your claw is solid, but the Sun's gaseous material allows it to behave differently and rotate at different rates."

The shimmering lights from the dragonflies' wings reveal momentarily eerie forms, things that look like eyes swimming and swiveling to stare at him with an incandescent glare, more like a rainbow than something alive.

"There are numerous ways of obtaining information about the Sun, including ones that focus on different wavelengths of light coming from the Sun. These different wavelengths are produced in regions of different temperatures. One of my favorite techniques employs the *spectroheliograph*, a device invented by American astronomer George Hale in 1890. This led to the discovery of magnetic fields and vortices in sunspots. It allows us to image the Sun in light of a single wavelength."

"How does the spectroheliograph work?" asks Miss Muxdröözol.

"It produces a photograph called a *spectroheliogram*. The device uses a prism, or grating, and a narrow, moving slit that passes only one wavelength of light to a photographic plate or digital detector. Brunhilde, show observation aids." The flexscreen lists various methods used to observe the Sun in great detail.

Table 6.2
How We Learn About the Sun

Observation Aid	Area of Study
Optical solar telescope (on Earth)	Sun's visible surface
Radio telescope arrays (on Earth)	Radio waves coming from the Sun
Infrared telescopes (on Earth)	Solar limb and sunspots
Spectroheliographs and color filters (on Earth)	Local phenomena; single wavelengths coming from Sun; single color light from a gas such as calcium or hydrogen.
UV, X-ray, and gamma ray telescopes (in space)	Processes in the Sun's hottest and most active areas.
Coronagraphs (on ground or in space)	Solar corona

The air seems a few degrees cooler. An odor, not quite perfume but more animal-like, permeates the air. Bob hands a card to Miss Muxdröözol. It reads:

🦋 *The Day the Solar Wind Disappeared* 🦋

"Tell me more," she says, obviously excited about the striking headline.

"For two days in May 1999, the solar wind that blows constantly from the Sun virtually disappeared. It was the most drastic and longest-lasting decrease ever observed."[12]

"And just what is the solar wind?"

"The solar winds are coronal gasses that include charged particles that stream off of the Sun at speeds of 400 kilometers per second (about 1 million miles per hour). The temperature of the corona is so high that the Sun's gravity doesn't hold on to all of it. In 1999, the wind dropped to a fraction of its normal density and to half its normal speed, which allowed physicists to observe particles flowing from the Sun's corona to Earth. This amazing change in the solar wind also changed the shape of Earth's magnetic field. Also, because of the decrease in solar wind, energetic electrons from the Sun were able to flow to Earth in narrow beams called the strahl."[13]

"Mr. Plex, put your polyehdral lips to the orange skin and blow through one of the small holes you have made. Yes, that's it. Your blowing is a metaphor for the solar wind, the stream of electrically charged particles that continually come out of the Sun. You can think of the solar wind as the corona expanding into the solar system."

"How far does the solar wind go?" asks Miss Muxdröözol.

"All the way to the *heliopause*, far past the orbit of Pluto."

Miss Muxdröözol comes closer. "That's an impressive sounding word, heliopause. We had a popular movie on my world, *Last Tango in the Heliopause*, starring a robotic version of Marlon Brando, but I never knew what the word meant."

Bob has Brunhilde display two words that oscillate in indigo and crimson and in various elaborate fonts. He shows the result to Miss Muxdröözol:

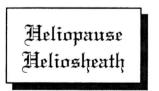

"Will you do anything to impress a lady?" she says.

"The heliopause is essentially the edge of the solar wind. Outside the heliopause lies the *heliosheath*, a region of transition between the *heliosphere* and the hydrogen atoms and galactic magnetic fields in outer space. Incidentally, the wind is a lot faster and thinner than a wind on Earth. Its average speed is 400 kilometers per second or about 800,000 miles per hour. The fastest part of the solar wind leaves the Sun through coronal holes, cool spots in the corona. The solar wind finds it easier to escape here because the magnetic field of the Sun is relatively weak around coronal holes. The Sun rotates as it emits the solar wind, so the solar wind spirals around the solar system."

"Sir, you mentioned heliosphere. What is it?"

Bob continues to walk through the grass as his friends follow. "Imagine the solar wind moving out into space like an expanding cloud. As it enlarges, the wind creates a magnetized bubble of hot plasma around the Sun, called the heliosphere. Eventually, the solar wind travels so far that it encounters the charged particles and magnetic fields of the faraway interstellar gasses. The boundary created between the solar wind and interstellar gas is the *heliopause*. We don't know the exact shape and location of the heliopause. Perhaps it extends 110–160 AU (astronomical units) from the Sun."[14]

Miss Muxdröözol grins. "Let me see if I have this all straight. This heliosphere is the big, big region that starts at the Sun's surface and extends beyond the solar system. This means that our solar system is filled with racing, charged particles from the Sun. The magnetic field carried by the particles creates a magnetic bubble called the heliosphere. "[15]

"You've got it! Another name for heliosphere is *magnetosphere*. Again, the heliopause is the boundary between the Sun's magnetosphere (i.e., the solar wind) and interstellar gas."

The air grows even cooler, but Bob is used to colder climates. He spent some of his adult years in Oswego, New York, where winters were bitter and the snow deep. Apparently Mr. Plex and Miss Muxdröözol don't seem to mind it either.

Bob withdraws a book of matches from his pocket and hands it to Mr. Plex. "Mr. Plex, I want you to poke a hole in a piece of rind of the orange and stick a lighted match through it."

"Yes Sir."

A flame pokes through the orange skin.

"The flame coming from the orange represents a solar *flare* that shoots from the Sun. Remember, I told you about them when we talked about twisted magnetic fields near the Sun. A huge flare on the sun releases more energy than the planet Earth uses in 100,000 years.[16] Just like your match, most solar flares only live for a short time, some only a few minutes. But very large flares can last over an hour."

Miss Muxdröözol stares at the flame. "Is a flare just visible light?"

Bob shakes his head. "A flare emits radiation across a huge range of the electromagnetic spectrum, from radio waves, at the long wavelength end through optical emission to X-rays and gamma rays at the short wavelength end. The amount of energy released is the equivalent of millions of 100-megaton hydrogen bombs exploding at the same time!"[17]

Now Mr. Plex stares at Miss Muxdröözol's white body and retractable canines. "May I remind the lady that a kiloton is equivalent to blasting 1,000 tons of TNT, and a *megaton* has a blast equivalent of one million tons. As comparison, the bomb humans dropped on Hiroshima, Japan, in 1945, released energy equaling 15,000 tons (or 15 kilotons)."

Bob nods. "Humans didn't discover solar flares until 1858. It was at this time that scientists were studying sunspots and suddenly saw a large flare of white light. In fact, solar flares occur near sunspots, and the number of solar flares rises and falls with the 11 year cycle of sunspots that I told you about. Solar flares appear to be triggered by strong magnetic fields."[18]

Miss Muxdröözol points to the two original dragonflies that float away like tiny bright kites towards a swamp. They oscillate all the colors of the rainbow, hover, and then begin to move rapidly. Miss Muxdröözol giggles nervously as she watches the creatures bounce from cattail to cattail like luminous balls in a Ping-Pong game. They go faster and faster and then spiral soundlessly down into the duckweed.

"We humans like to study the solar flares because it takes a flare just eight minutes to affect the Earth's atmosphere. The high-energy radiation and electrically charged particles can cause the Earth's upper atmosphere to becomes more ionized and to expand, which can mess up long distance radio transmissions, damage satellite electronic components, and even harm astronauts.[19] During maximum solar activity, as I said, the Earth's upper atmosphere can heat up by a factor of three, causing the atmosphere to expand. As a result, satellites in low Earth orbit can experience increased drag and reenter the atmosphere. All life on Earth would be wiped out if our atmosphere and magnetic fields did not protect us" (figure 6.9).

Bob sees that Miss Muxdröözol sometimes shivers. "If it doesn't get warmer soon, we can try to build a fire." The thin cotton fabric of her shirt barely hides her tiny goose bumps that seem to travel in concentric waves from her neck down.

"What are you staring at?" Miss Muxdröözol says.

"Nothing," Bob returns to his Sun lecture. "Even with our protective atmosphere, particles from the Sun can interact with the Earth's magnetic field and cause magnetic storms in which compasses don't work. Particles can also trigger power blackouts when the magnetic storms cause currents to flow in long power lines. The currents blow circuit breakers. The beautiful auroral glows— the northern and southern lights—are triggered when the solar wind brings

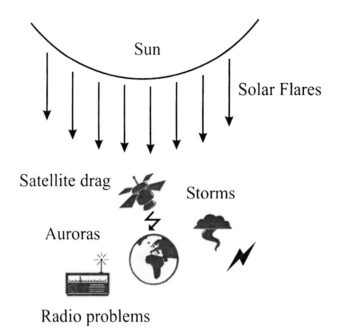

Figure 6.9   Solar mischief. Solar flares can cause the Earth's atmosphere to puff up like a toasting marshmallow, leading to additional drag on Earth-orbiting satellites. Solar flares can also disrupt communications, cause storms and auroras, and disrupt the Earth's magnetic fields so compasses don't work.

high-energy atomic particles that emanated from sunspots. Solar flares are so dangerous that we try to predict them, but it's very difficult to foretell exactly when a flare will occur."

Bob walks over to a large juniper tree. "Wonder what this is all about," he says pointing to the tree. He bends lower. Someone, maybe a child, has carved the word *Zôrôêl* into the thick trunk of the giant juniper. The deep gashes have weathered so that they are a shiny beige against a rough, peeling bark. *What a strange word*, Bob thinks.

"Obviously not English," says Mr. Plex.

Bob nods. "*Prominences* are another type of eruption from the Sun. They look like *arches* of fire and are made of ionized gases. Prominences can last for many weeks. They also seem to be caused by magnetic fields near sunspots. They can rise over 10,000 kilometers from the Sun" (figure 6.10).

"Huge," says Miss Muxdröözol. "The Earth would easily fit under one of the loops of the prominence."

Bob nods. "Some prominences erupt, spewing enormous amounts of solar material into space. The most violent of solar eruptions is a *coronal mass ejec-*

Figure 6.10   Sun image showing a portion of a large, eruptive prominence. The image was
taken by a SOHO (Solar and Heliospheric Observatory) spacecraft. The SOHO project is
being carried out by the European Space Agency (ESA) and the U.S. National Aeronautics
and Space Administration (NASA) in the framework of the Solar Terrestrial Science Program
(STSP). (Figure courtesy of the SOHO/EIT consortium and NASA.)

tion (or CME). These are large, balloon-shaped plasma bursts that come from
the Sun. They move along the Sun's magnetic field lines and can be up to tens
of millions of degrees. The bursts release up to 220 billion pounds (100 billion
kg) of plasma. Just like with solar flares, CMEs can cause Earth's satellites to
fail. CMEs are sometimes associated with solar flares" (figures 6.11 and 6.12).

Suddenly, Bob sees the large ants coming toward them. Their faceted eyes
glow like embers in a dying funeral pyre.

The larger one with the golden abdomen says, "I hear them." Its voice sounds
as if it is warped by some kind of artificial translation device. Again, it twists an
electrical probe up its huge nostril.

The creatures turn and amble closer. Bob sees that their bodies are actually
suspended from their long arched legs.

The smaller ant hisses with an aggressive machine-gun speech. "Right, it
must be them. Let's get them." Its hissing turns to a gurgling grunt as it gestures
toward Bob and Miss Muxdröözol with two of its right legs. Mr. Plex has already dived into the grass for cover.

As the ants come closer, there becomes a more stalking, purposeful intent in
their movements. Mr. Plex starts to cough. The ammoniac smell of urea becomes overpowering.

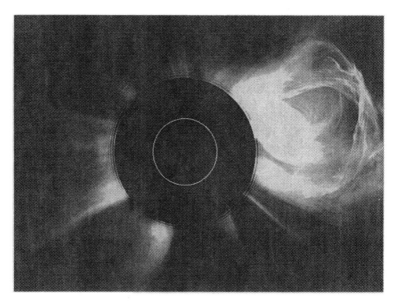

Figure 6.11 Coronal mass ejection. To produce this image, direct light from the Sun is blocked at center, but the sun's position and size are indicated by the central circle. [Figure courtesy of: NASA, the Solar and Heliospheric Observatory (SOHO Consortium), and the European Space Agency (ESA). See also "Astronomy Picture of the Day," http://antwrp.gsfc.nasa.gov/apod/ap000309.html]

Figure 6.12 Dramatic high-resolution image looking across the edge of the Sun. Shown here are arcs of hot gas suspended in powerful looping magnetic fields that rise above a solar active region. The image was made in the extreme ultraviolet light. [Figure courtesy of CFA (a solar and stellar X-ray group at Harvard University), TRACE (Transition Region and Coronal Explorer) Team, and NASA.]

"Run for it," Bob says and pulls on Miss Muxdröözol's vestigial arm. Mr. Plex is close behind.

"Sir, they're gaining on us," Mr. Plex pants, looking back at their pursuers. Bob feels the violet grass slap against his legs like soft whips.

Miss Muxdröözol trips and falls, crashing through the grass. Her legs are violet with exudate from the thin stalks.

"Forget me," she cries to Bob. "Go save yourself."

Bob stops and reaches for her. "Come on!"

The creatures arrive in a few seconds, stop, and appear to sneer at Bob. Their single nostrils are pinched. They breathe rapidly, evidently exhausted from running.

The smaller creature turns his deep-set black eyes on Miss Muxdröözol's. Her body tightens and she takes a deep breath. "God, what do we do now?" she says. She picks herself off the ground and is seized by the small ant with the colorful blinking abdomen. Its exoskeleton has crescent-shaped scales crisscrossed with fine lines. Its legs are gaunt, each tendon clearly evident beneath the dark, hard skin. Its fangs, pushing delicately against its lower jaw, have the appearance of ebony.

Bob tries to muster as much bravado as he can. "Take your hands off my wife, you oaf."

"Silence!" hisses the larger ant. It came out *sssssilensss . . .*

The smaller one unfolds a sitting-stalk from its abdomen and perches as if waiting for a command from the other.

"Listen buddy," Bob says, "no one touches my wife that way. Don't you realize she's an undercover agent? She's bringing rare metals to your queen."

"What?" the creature says as it wrinkles its brow. "What are you talking about?" The other ant makes crackling sounds.

"These," Bob says. He slowly brings his hand from his jacket pocket. In it is a handful of change, mostly quarters and shiny pennies. Bob shakes his hand so the coins catch some of the ambient light and make a clinking sound.

"Let me see those," the creature says, jerking its huge legs like a marionette controlled with strings from above.

Bob is terrified but conceals his emotions as one of his faces fixes on the larger alien while his other face watches the ant holding Miss Muxdröözol. The bluff is going well, he thinks. All he has to do is throw the coins in the ants' faces and then run with Miss Muxdröözol. He feels a rush of adrenaline as he clenches his hand, ready to hurl the money. Mr. Plex remains inscrutable, motionless like a statue.

"Wait," Miss Muxdröözol says.

Bob wonders what she is doing.

"Never mind the precious metals," she says. "Show him the nuclear-powered writing instrument we carry as a gift to their leader." Her face is utterly serious.

"She's right," Bob says, not knowing what in the hell she is talking about. She reaches into his pocket and brings out an ordinary Bic ballpoint pen.

"You see this?" she says. "Watch." She starts drawing zigzagging lines on her palm:

Does she really think that they would be impressed with a pen? The larger creature comes over and tries to grab the pen from her hand.

"I want that," it says. The two jerking creatures circle her, shuffling and bobbing along the ground.

"Wait just a minute," Bob says. "Before we let you hold these gifts, how can we be sure you will not steal them and kill us."

"We have to teach you to use them," Miss Muxdröözol says. "The pen's operation is really quite complex."

Bob sees that her face betrays a certain tension held rigidly under control. Even Mr. Plex seems to be clenching his inorganic teeth. *Could an alien ant understand human facial expressions?*

The creature snarls and reaches for Miss Muxdröözol's pen. "Just give them to me."

Miss Muxdröözol is quick. She closes her hand around the pen and then throws it at the creature's face. She must have been an excellent marksman because it struck the ant right in the face. Its abdomen pulsed all the colors of the rainbow.

"No," the creature says, clutching its face.

Bob throws his coins at the other ant. "Run," Bob shouts to Miss Muxdröözol and Mr. Plex. "C'mon, get going."

Fear and adrenaline propel them forward with such speed that soon the ants are about 40 feet behind them.

"We're lucky they seem so out of shape," Miss Muxdröözol gasps. "They can't keep up with us." The ground starts to shift beneath her feet.

Bob pulls her hand. She regains her balance but then stumbles again when the loose soil shifts.

"Watch out for the hole!" Bob says.

Miss Muxdröözol screams as the ground begins to crumble. She avoids falling into the dark oubliette only by a few inches.

"Help," she says as Mr. Plex races to grab her. But he loses his footing and sprawls headlong into a bush.

Bob reaches out to steady her, but they both lose their balance. Hand in hand, they fall into the hole. They slide along the smooth dirt wall and finally land in an emerald green liquid encrusted with a rust-colored powder. They hit the liquid feet first and quickly sink in up to their waists. Bob thinks he feels something move beneath his feet. Is it alive? Or is it just sludge shifting beneath his weight?

"Arrgh," Bob cries. "You okay?"

"Yeah," Miss Muxdröözol says as she tries to hold herself back from vomiting. A piece of viscous green goo on her hair slides down her forehead like some horrible slug.

"Let's get out of here," she says, apparently trying to discipline her voice, to maintain complete control.

Again Bob thinks he feels something move under his feet. He sinks deeper into the hole, up to his chin. Slimy and squishy matter presses all around him.

"Help," he cries. Bile-green, stinking surface scum washes into his mouth. The smell of Limburger cheese and rotten eggs is everywhere. "I'm going to drown!" he says. He would have vomited if he had any substantial amount of food in his stomach.

"Sir, take my claw," Mr. Plex says, holding onto some protruding tree roots with his free claw. "I'll pull."

In the next few minutes they are able to slowly climb up the wall of the hole as they slip, slide, and curse. When they finally emerge from the hole they begin running again. The wet muck flys from their bodies like sparrows fleeing a nest of snakes.

"No!" Miss Muxdröözol cries when she sees that the ants are only about 20 feet away from them.

"Keep running," is the only thing Bob manages to say. His chest begins to ache as he breathes hard.

"Wait," cries one of the ants. "We only want to take you to the Nephilim."

And with those words Miss Muxdröözol explodes, her internal organs shooting outward into chasms of fetid air.

Some Science Behind the Science Fiction

All scientific knowledge that we have of this world, or will ever have, is as an island in the sea of mystery. We live in our partial knowledge as the Dutch live on polders claimed from the sea. We dike and fill. We dredge up soil from the bed of mystery and build

ourselves room to grow . . . Scratch the surface of knowledge and
mystery bubbles up like a spring. And occasionally, at certain
disquieting moments in history (Aristarchus, Galileo, Planck,
Einstein), a tempest of mystery comes rolling in from the sea and
overwhelms our efforts . . .

— Chet Raymo, *Honey From Stone: A Naturalist's Search for God*

Bob mentioned how shock waves from a nearby supernova, or exploding star, may trigger the gravitational collapse of interstellar gas to form solar systems. In our Galaxy, a star becomes a supernova about once every 20 years, although there is a large uncertainly to this number. There are about 100 billion galaxies within the observable Universe. This suggests that the observable Universe experiences 100 supernovas every second. More than half a billion stars have detonated since the Milky Way was born.[20]

Supernovas and novas are both rapid brightenings that last several weeks. However, *super*novas, unlike novas, are events that end the star's energy-generating life. When a star becomes a supernova, huge amounts of matter are shot into space, and this matter is equivalent to the mass of several suns. During the explosion, the star may outshine an entire galaxy. Not only do supernovas create new elements, but they project those elements into space along with many of the elements forged in the normal course of a star's life. Whereas red giants eject mostly medium-weight elements such as carbon and oxygen, supernovae alone shoot significant quantities of the very heaviest elements into space.[21]

On the other hand, most, if not all, ordinary novas occur in double-star systems. For example, consider a red giant in the vicinity of a white dwarf. The gravitational field of the white dwarf may pull hydrogen from its larger companion, thereby initiating fusion and causing a nuclear explosion, a nova, that launches a small amount of gas into space. The process may repeat many times.

As I just mentioned, one to two billion stars have exploded since our Galaxy formed.[22] I wonder if life formed on planets around some of the stars before the stars became supernovas. I wonder if the creatures knew that their deaths were essential for scattering heavy elements into the vast empty reaches of interstellar space. In time, these elements congealed into new stars, and the cycle of birth and death occurred many times since the birth of our Galaxy. Our Sun is probably a third-generation star, meaning that two generations of stars had to self-destruct for humans to be born.[23] Without the death of earlier stars, we would not have our rocky planets like the Earth and Mars, and there would be no heavy elements like carbon, nitrogen, and phosphorous for biochemical reactions in cells. But even the infant Milky Way was not totally devoid of heavy elements. Scientists feel that our Galaxy contained some heavy elements when it was born because no star ever studied has been without heavy elements. The first heavy elements in the Milky Way probably came from stars that lived and died before galactic formation, perhaps in the first billion years of the Universe.[24] We can only speculate about what these ancient stars were like and how they came to light up in the baby Universe. Marcus Chown, author of *The Magic Furnace*, suggests that soon after the big bang, clumps of gas began to congeal, and then the Universe began to light up like a Christmas tree.[25] I call these stars "Genesis stars" because they lived and died before our Galaxy came into existence, eons before the Age of Man. I don't know if any pre-galactic stars are viewable from Earth today.

* * *

Bob and Mr. Plex discussed sunspots and the effect of solar output on the Earth's climate. This is obviously an important topic with dire consequences. From 1645 to 1715, for reasons not yet known, almost no sunspots were observed. It coincided with a period of colder-than-average temperatures in northern Europe. As Bob mentioned, astronomers call this period the Maunder minimum.[26] Historians refer to the period as the Little Ice Age, the coldest period of time since the last Great Ice Age 10,000 years ago. During the Little Ice Age, Scandinavian and Swiss glaciers extended down mountain valleys, destroying farms. Rivers in the southern United States froze, and the Thames River in England froze for the first time in history. Europe's population stalled, crops failed, and grain prices skyrocketed. Sea ice cut Iceland off from the mainland. Aside from actual reports of the lack of sunspots, changes in the Sun's past activity can be gleaned from the amount of radioactive carbon-14 in tree rings, which suggest that solar radiation decreased by a quarter of a percent during that period. Just a quarter of a percent changes the world. What will happen to our civilization when the sunspot cycle disappears again, as most astronomers predict it surely will?

The surface of the Sun today has an effective temperature of 5,780 K, but five billion years ago, it was cooler, about 5,500 K. The Sun was also smaller and produced 70 percent of its present radiation. Today, a 30 percent decrease in solar luminosity would destroy our ecosystems. Water would freeze, and the planet would become more like Mars. However, perhaps early Earth could have supported life because the ancient atmosphere had more carbon dioxide and therefore trapped more solar heat.[27]

* * *

When thinking about how our solar system may have evolved from "proplyds" (protoplanetary disks), we must remember that the violence of the early Solar System was tremendous as huge chunks of matter bombarded each other. In the inner Solar System, the Sun's heat drove away the lighter-weight elements and materials, leaving Mercury, Venus, Earth, and Mars behind. In the outer part of the system, the solar nebulas (gas and dust) survived for some time and were accumulated by Jupiter, Saturn, Uranus, and Neptune.

On religion I tend toward deism but consider its proof largely a problem in astrophysics. The existence of a cosmological God who created the universe (as envisaged by deism) is possible, and may eventually be settled, perhaps by forms of material evidence not yet imagined.

— E. O. Wilson,
Consilience: The Unity of Knowledge

CHAPTER 7

Stellar Evolution and the Helium Flash

When the stars threw down their spears
And water'd heaven with their tears,
Did he smile his work to see?
Did he who made the Lamb make thee?

— William Blake

"**S**ir, there's a leech on your left arm."

"It's okay, Mr. Plex. This is one of the new electronic ones. I apply them occasionally, and they dump free-radical scavengers into the blood. I'm trying to regain some energy. But, my God, what happened to Miss Muxdröözol?"

An ant approaches, "Do not worry, your friend is perfectly safe. For the last five minutes you have been watching a holographic representation of the entity you call 'Miss Muxdröözol.'"

Bob steps closer. "How do you know us? Where is she?"

"Please forgive us for our initial confrontation. We wanted to test your stamina and ingenuity. We are, in fact, a peaceful race, keepers of the veined membranes, the wormholes between space. We've been watching you since you left your art museum. The iceworms that you threw onto your green canvases are actually our eyes. You might call them remote viewing devices."

The ant with the multicolored, blinking abdomen nods. "We'll bring Miss Muxdröözol to you shortly. We want to take you to see the Nephilim."

"Nephilim?" Bob says.

The ant remains motionless, and its blink rate slows.

Mr. Plex claps Bob's shoulder with his feeding appendage. "Sir, the most enigmatic of biblical stories deals with the Nephilim. This strange race is mentioned in Genesis 6:1–4."

Mr. Plex is amazing, but sometimes Bob is uncomfortable with the kind of creatures who could go on *Jeopardy* and win or could quote any passage from the Bible. However, this was not a time to shun any information Mr. Plex could give. "You know our Bible, Mr. Plex?"

Mr. Plex nods. "Yes, most of the scolexes on my world have been Christianized by the metallic missionaries. We devote half of our waking hours to studying the Bible, particularly the Old Testament.

Bob recalls rumors that several Christian and Jewish sects had sprayed tiny message-carrying robots into the cosmos. These messengers went to orbit stars and awaited the possible awakening of civilizations on nearby planets. Because a messenger may have had to wait millions of years before making contact, it was heavily armored to withstand radiation damage and meteorite impacts. The messengers also were said to have powers of self-repair and replication and to get their energy from starlight.

The ants remain inscrutable as ever.

Mr. Plex continues. "In Genesis 6:2 we find that the 'sons of God saw that daughters of men were beautiful, and they married any of them they chose.' Religious scholars speculate that the 'sons of god' might have been angels who took wives from the daughters of humans. The offspring of these angel-human marriages were the Nephilim, the 'heroes that were of old, warriors of renown.'"

The ants seem to nod, but it's difficult to tell given their strange anatomy.

Bob puts his hands on his hips. "We've got to find Miss Muxdröözol."

Mr. Plex interrupts. "The Nephilim are mentioned only once again in the Hebrew scriptures, and the word also literally translates to 'the fallen ones.' The Nephilim had superhuman powers. Notice that they should have been destroyed in the great Flood, but we do find them in Canaan during the time of Moses, according to the book of Numbers."

Before Bob has a chance to fully assimilate this deluge of biblical information he hears a rumbling noise to his right, and soon a large chariot pulled by eight ants comes beside him. The chariot is enclosed and has small windows and a transparent sunroof.

"Please enter," says the ant. "We will not harm you. We wish to take you to the Nephilim."

"Wait," Bob says. "What are your names?"

"Call me Ishmael," says the ant with the golden abdomen.

"I am Cain," says the other ant. "I will help you." Cain bows while his blinking abdomen pauses on red for several seconds before resuming its red-green oscillations.

Mr. Plex enters the unusual vehicle. Bob hesitates but then follows.

In the back seat is Miss Muxdröözol wearing a beige, double-breasted, French hand-finished blazer with narrow-legged matching trousers. Beneath she wears a soft, saffron silk blouse with a high cossack collar and pearl buttons.

She smiles. "Don't worry, Bob, I'm okay. The ants gave me these clothes."

The air inside the chariot has a slight marine smell, as if the chariot has been near some coastal shore. After a few seconds, the chariot seems to ride almost effortlessly. Perhaps it is the low gravity.

"Look up!" Mr. Plex says. "What happened to all the stars?"

Bob looks up. Is that still the Milky Way Galaxy that was previously overhead? He can make out a bright central region of stars. There are patches of red. The spiral arms are gone. "The stars are dying, Mr. Plex. Evidently, we've moved again further into the future. Remember, when we left Earth, our Sun was five billion years old, halfway through its life."

Miss Muxdröözol nods. "The sun must have died by now." She pauses, "It must have become a red giant, hundreds of times its original size. The inner planets were burned to cinders." Her voice takes on a sad timbre. "The land was hot enough to melt lead."

"Yes," Bob says, "and beyond that the smallest, longest-lived dwarfs can last 100 billion years, but sooner or later there will be no more new stars. Then all we'll have are the dense black holes, neutron stars, and some white dwarfs. Maybe finally everything will collapse into huge black holes."

Bob looks up again. "It all seems bleak. The sky is simpler now—galaxies must have receded far from one another. Yet, there is beauty in the colors. Perhaps some of the stars are glowing with colors never seen in the young Universe as they become rich in metals forged in the generations of suns that have been born and died.

"Sir, my instruments indicate that the Big Bang glow is 0.01 degrees above absolute zero. The Universe's temperature decreases as the two-thirds power of time.[1] That means we're about 10^{14} years into the future!"

Miss Muxdröözol puts her head into her hands. "It's so sad to see the Galaxy die, and with it, all life."

Bob studies the sky. "Not so. Life didn't die. Look over at some of those bright stars. They're not natural."

The ants grin. "The Nephilim."

"What?" says Mr. Plex.

"They can gather interstellar material together and form artificial stars."

Miss Muxdröözol looks at Bob and the ants. "What do you mean? You mean they can create new stars?"

Bob nods. "Apparently so." He pauses. "Today I will teach you about *stellar evolution*. This will help us all understand what is happening and what might happen next."

The ants nod in unison. "Tell her about the stars."

Bob leans back in the Naugahyde chariot seat. "Stellar evolution refers to what happens to stars as they get older. We can use the H-R diagrams I already told you about to visualize this. In our twenty-first century, stars were born all the time in huge clouds of gas and dust called nebulae. Think of these clouds as stellar nurseries. One of the most famous is the Orion Nebula" (figure 5.5).

"Right," Miss Muxdröözol says, "you told us about the Orion proplyds. They were larger than the Sun's Solar System and contained enough gas and dust to provide the raw material for future planetary systems."

"Yes, fetal stars, or *protostars*, form when hydrogen gases and dust clouds coalesce in space. Gravity holds a protostar together, and as the protostar contracts, temperature and pressure start to rise. The fetal star begins to shine in the infrared. When the temperature in the center of the fetal star rises to ten million degrees Kelvin, a mature star is born, and nuclear fusion reactions start. The outward force of the intensely hot gases counteracts the inward crushing force of gravity to maintain a *hydrostatic equilibrium*" (figure 7.1).

Bob looks out the window at the changing landscape. They are traveling on a dirt road. Trees are becoming more common, and soon forests beside the

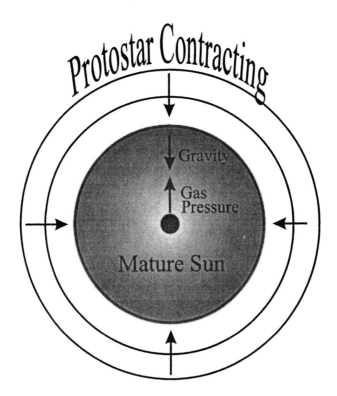

Figure 7.1   The force exerted by the hot gas counteracts the crushing force of gravity.

road are dark with mystery. Bob sees streams, and the ants permit Bob and his friends to stop and get a drink.

Bob washes his faces and hands and returns to the chariot. "This solar equilibrium almost seems to be a miracle. If gravity were much stronger or much weaker, this balance between gravitational force and the outward pressure of gas would be disturbed. For example, if the gravity was off by just a little bit, nature would tend to produce more extreme stars—hot blue giants or cool, dim red dwarfs— neither of which seem very conducive to the evolution of life."[2]

"The formation of yellow suns depend on the gravitational constant," Mr. Plex says.

"Yes, if it were significantly different, every star in the Universe might be either a blue giant or red dwarf for the outward gas pressure to balance the gravitation.[3] No yellow suns would exist to warm Earthlike worlds. Remember, nuclear fusion is the same process that provides the tremendous energy for hydrogen bombs. Luckily, the Sun's fusion is self-regulating in the sense that the inward pull of gravity is counteracted by the fusion energy. If there was no nuclear fusion, the Sun would have contracted into a hot, dense white dwarf in just a few million years."

The fields on either side of the road are dotted with thickets of amber and ochre ghostlike trees. The road becomes narrower.

"Let's talk about the varieties of stars in the heavens. In a nutshell, stars are different because the gas and dust clouds from which they emerge have different masses and densities. However, if two stars emerge with the same initial mass and chemical composition, their evolution and eventual fates will be the same."

Miss Muxdröözol nods. "Just like our genes determine our own development, the initial makeup of stars determines theirs."

Bob nods, one face looking at Miss Muxdröözol, the other at Mr. Plex. The ants just stare in silence. "Low-mass stars take the longest time to evolve. High-mass stars evolve quickly, at least on stellar timescales."

"What's an example?" asks Mr. Plex.

"Well, our Sun takes about 30 million years to incubate as a protostar and form a mature sun. Stars three times the mass of our Sun might take just a million years to be 'born,' and stars one tenth the size of our Sun might emerge in about 100 million years. Protostars are just dense clouds of gas and dust. Remember, a supernova triggers their gravitational collapse into a star" (figures 7.2 and 7.3).

Bob looks out the window and sees his first waterfall, although it behaves rather strangely in the lower gravity. The water seems to be falling more slowly than normal.

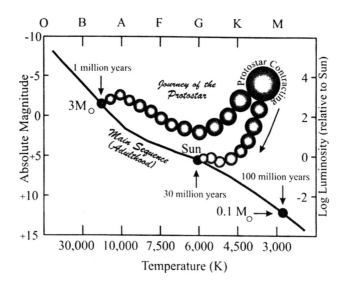

Figure 7.2 Protostars evolve into mature stars located on the Main Sequence. At upper right is a contracting protostar. Stars of different masses have different fates and "travel" different paths along the luminosity versus temperature plot. The fates of stars (represented as solid circles •) with masses three times the mass of the Sun, equal to the Sun, and a tenth of a Sun's mass are shown. (For diagrammatic clarity, only two paths from protostar to mature star are depicted.)

Figure 7.3 Molecular gas clouds in the Eagle Nebula (M16) star-formation complex. This wonderful Hubble Space Telescope image shows regions of molecular gas and dust, most with faint embedded protostars, that are illuminated by a hot young star outside of the frame of the image. The individual lumps breaking off the ends of the three long "fingers" of gas are about the size and mass of the pre-stellar clumps from which stars like the Sun might have formed. In this image, however, many of these clumps are probably being evaporated by the hot UV radiation from the nearby hot stars before they have a chance to form into low-mass stars. (Figure courtesy of NASA and AURA/Space Telescope Science Institute.)

"You can think of all the stars on the Main Sequence as mature stars, but the story doesn't end here. There is additional, although much slower, evolution of the star after it reaches the Main Sequence. Once a star is on the Main Sequence, it produces energy from nuclear fusion reactions in which four hydrogen atoms are fused into a lighter, helium nucleus. It's really the separation of hydrogen nuclei from their electrons, and the extreme temperatures, that make nuclear fusion possible at the Sun's core. Brunhilde, display fusion diagram." Brunhilde displays figure 7.4.

"Separation of nuclei from their electrons?" says Miss Muxdröözol.

"With their electrons gone, hydrogen nuclei (or protons) can be squeezed together very tightly. Deep inside the Sun, the pressure of surrounding material is huge, causing the nuclei to pack tightly. Remember, at the Sun's center, the temperature is 15.6 million degrees Celsius (28.1 million degrees Fahrenheit), and the density is more than 14 times that of solid lead. This extreme heat and density fuses the nuclei together."

Mr. Plex watches some birds, soaring overhead, calling. The calls seem unnatural, almost as if they were produced by an electronic oscillator. But who can tell what "natural" means in a world like this. "Sir, you said that four hydrogen nuclei are fused into one helium nucleus. But where did the extra protons go. Your plot shows helium with just two protons."

"Two of the original protons become neutrons—as you know, they're electrically neutral particles about the same mass as protons. The result is a helium

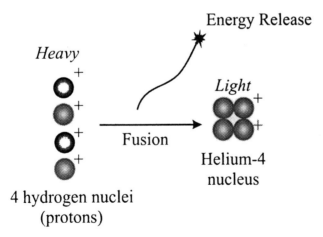

Figure 7.4   Why stars shine. Energy is produced when four hydrogen nuclei fuse into a helium-4 nucleus. (There are six known isotopes of helium, but only two are stable: helium-3 [symbolized ³He] and helium-4 [symbolized ⁴He], which is the most plentiful of the stable isotopes.)

nucleus, containing two protons and two neutrons. The resulting helium nucleus is 0.7 percent less massive than the four protons that combine to make it. The fusion reaction turns the missing mass into energy that allows the Sun to heat the Earth and permits the formation of life. The mass loss, m, during the conversion of four protons into one helium nucleus, supplies an energy, E, according to Einstein's famous relation $E = mc^2$, where c is the speed of light (3×10^8 meters per second or 1×10^9 feet per second). Every second, fusion reactions convert about 700 million metric tons of hydrogen into helium within the Sun's core, thereby releasing tremendous energy. And that is why, Mr. Plex, the Sun shines."

"But wait," says Miss Muxdröözol, "why doesn't the Sun just disappear if so much mass is converted to energy?"

"On solar scales, the change is slow. It takes a billion years for only 0.01 percent of the Sun's mass to metamorphose into beautiful sunshine. The Sun's nuclear reactions are slowed because the positively charged protons repel each other. This repulsion slows down the fusion. If the rate were much quicker, the Sun would explode like a big hydrogen bomb."

"But eventually the Sun uses up all its hydrogen?" says Miss Muxdröözol.

"Yes, eventually all the hydrogen in this case is converted into helium."

Miss Muxdröözol looks up at the sky as a shiver runs up her hyperbolically cantilevered back. "It's sad. The Sun must have died by now."

"Yes, the light, cool, dim stars lived the longest, billions of years, because they burn their hydrogen slowly. The red dwarfs have little mass and are the oldest and most numerous Main Sequence stars. Think of them as stingy misers that hoard all their riches, spending as little as possible to maintain their lives. On the other hand, the massive, big, hot, bright stars die fastest because they burn their hydrogen rapidly. Blue giant stars might only stay on the Main Sequence for a few million years before they start to die. A mere blink of the eye."

"You talked about *red giants* before," says Miss Muxdröözol. "What else can you tell us?"

"Remember when I told you about Betelgeuse, the red giant superstar? That was a particularly big red giant, but our Sun also eventually turned into a red giant when it used up all the hydrogen in its core." Bob looks out the window, "Imagine a dying star. It has devoured the hydrogen in its core, so the core begins shrinking, which converts hydrogen into helium in regions that were previously surrounding the core."

"Does the star shrink because the hydrogen is being used up?"

"The outward pressure of heat generated by the nuclear reactions no longer balances the inward gravitational attraction."

"But wait," Miss Muxdröözol says, "how does a *shrinking* star create such a giant?"

"Well, it is the core that's shrinking as it fuses its hydrogen supply. When a red giant forms, gravitational contraction causes the core temperature to rise, and the star brightens. The new hydrogen fusion and contraction release so much energy that the star grows huge, like a balloon inflated by a manic blower. It's the outer parts of the star that expand so that the star's density is tiny except at the core. Eventually, the star expands into a red giant, with diameters about 100 times the diameter the Sun had in the twenty-first century."

"Sir," Mr. Plex says, "how did it all end on Earth?"

"When it was a red giant, the Sun had expanded to a size about as big as Earth's orbit and was 2,000 times brighter than normal. It's sad to imagine that the giant Sun turned all of humans' buildings to ash and oceans to steam. The churches melted, the Eiffel tower melted, the great pyramids melted, the Empire State building melted, the rocks melted—they all turned to a crisp, like the black ash on a toasted marshmallow."

"Sir, you're not making Miss Muxdröözol feel any better with that kind of talk."

"Actually, the huge, aging star's surface temperature begins to drop as its color turns red."

"How is it both cool and so bright?" asks Miss Muxdröözol.

"It's so bright because of its very large surface area. Take a look at the H-R diagram in figure 5.1, which shows you where the red giants are in terms of temperature and luminosity."

"Sir, why do you say 'figure 5.1'?" says Mr. Plex. "You're acting like we're in some kind of book."

Miss Muxdröözol looks up. "Can we see Betelgeuse now?"

"It's dead and gone, but I loved her as a man loves a woman. When I was on Earth, I could see Betelgeuse with the naked eye. Remember it was over 400 times the Sun's diameter. Although rare, the most massive stars can evolve into these stars called supergiants."

Miss Muxdröözol looks down at her feet. "It seems so pointless. All the stars die!"

"Apparently these so-called Nephilim seem to be able to rebuild stars," Bob says. "We have to find out more about that."

Mr. Plex says, "I recall your Nobel Prize-winning physicist Steven Weinberg once wrote, 'The Universe faces a future extinction of endless cold or intolerable heat. The more the Universe seems comprehensible, the more it also seems pointless.'"[4]

"Hold on, people. Don't be so morose. Now it's you, Mr. Plex, who is depressing Miss Muxdröözol. I told you that inside the larger red giants, fusion reactions can build up elements heavier than carbon; these elements include oxygen, aluminum, and calcium. We need stars to burn, die, and blow up to distribute the heavy atoms into the cosmos. Life needs the heavy elements of carbon, oxygen, and so forth."

The road on which the chariot rides is barely discernible amidst the grass. The chariot slows.

"Let me explain," Bob says. "The gravitational contraction of stars as they die causes the temperature inside the red giant's helium core to rise to 100 million K. That's wonderful, because at that temperature, helium is converted to carbon in nuclear fusion reactions. Now we're not talking about hydrogen fusion but *helium fusion*. Once the helium fusion starts, the helium core does not expand much, but the temperature rises without a cooling, stabilizing expansion.[5] The helium nuclei fuse increasingly fast. The core gets hotter. And hotter. And hotter. Bang. In low-mass stars the onset of helium fusion can be very rapid, producing a burst of energy called the *helium flash*. Finally, after a 100 million years of contraction, the helium core has contracted enough to ignite, which it does explosively over the course of a few hours!" (Figures 7.5 and 7.6)

"Sir, you mean that this helium flash occurs within a few hours after helium fusion begins?"

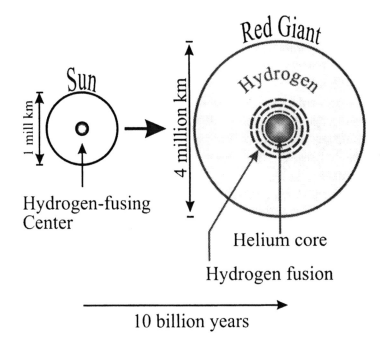

Figure 7.5   Our Sun, like similar stars, will become a red giant in 10 billion years. At right, the hydrogen in the core has been converted to helium. A hydrogen-fusing shell surrounds the helium core. [After Dina Moché, *Astronomy* (New York: John Wiley & Sons, 1998).]

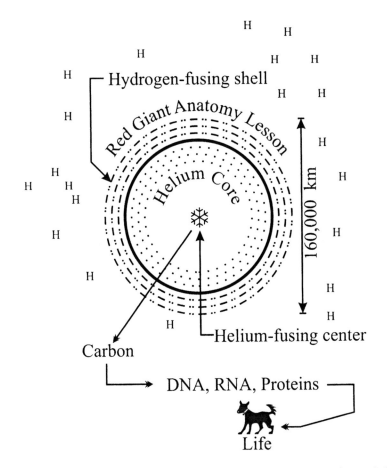

H H
H
H H H
H
H ┌ Hydrogen-fusing shell
H H
H
Red Giant Anatomy Lesson
H H
Helium Core
H H H
H H
H
Carbon
Helium-fusing center

160,000 km

Figure 7.6   Ancient stars have a carbon core surrounded by a helium-fusing shell
which is itself encased by a helium and a hydrogen-fusing shell. Small stars,
less than 1.4 times the Sun's mass, die quietly. Very massive stars exploded,
belching their carbon-heavy elements into space. These heavy elements, like carbon,
eventually found their way into primitive life-forms on Earth.

"Yes. Next, the temperature rises to a point that the core expands and cools. Helium fusion continues at a steady pace surrounded by a hydrogen-fusing shell. It then burns helium in the core for 50 million years, converting it into carbon. Think of a golf ball as a model for a helium core. The very center is helium fusing. Its dimpled surface is a hydrogen-fusing shell that is further surrounded by a hydrogen envelope."

Mr. Plex leans closer. "Sir, the helium flash—is it bright?"

"It is bright deep inside the star. The peak luminosity during the helium flash can be 10^{14} solar luminosities or about 100 times the entire energy output of our Galaxy. However, this huge amount of energy does not destroy the star.

The energy goes into expansion of the stellar core, and little of the energy reaches the surface. Once helium has been ignited in the core, whether explosively or not, the star has two sources of nuclear energy generation—helium burning in the core and hydrogen burning in a shell."

As the chariot rides on, the violet grasslands and forests give way to a meadow with tilting gravestones. It is an old cemetery with ornately carved stones. At the edge of the cemetery, where grass meets desert, are the oldest of stones, weathered to such a degree that their names are eroded away. Some of the graves are ruined, perhaps dug up by rodents or the large ants or destroyed by vandalism or the harsh winds.

Ishmael, the ant with the golden abdomen, begins to pulse his abdomen on and off. "Do not worry about the cemetery. It's there largely to keep out intruders, particularly those species that have an especially heightened fear of death, dying, and funerary rituals."

Bob looks at the headstones. This asteroid is so confusing that it makes his mind feel numb.

Bob continues. "The chemical makeup of a star depends on the distance from the star's center. As I've suggested, in the core are the heaviest elements, like carbon and oxygen produced by the helium burning. As the Sun collapsed, the pressure and temperature rose until it was high enough for helium to fuse into carbon. As you examine material further from the core, you find helium and then hydrogen untouched by nuclear reactions."

Bob looks out the window. Here and there are a few rotting coffins, and *Oh my God*, a few bleached bones are sticking up through the fractured wood of the coffins. Is that a robotic owl Bob sees picking on a few pieces of hair and splinters of bone?

Bob ignores the frightening sight. "Later I want to talk to you about the very final stages of a star's life, but for now just keep in mind that stars can move back and forth between the red giant region of the H-R diagram and the Main Sequence several times before they die. Some stars vary in brightness because they pulsate in and out. In the outer layers of pulsating stars, the inward pull of gravity and the outward push of pressure are out of balance. When outward pressure overwhelms inward gravity, the star expands. When inward gravity surpasses outward pressure, the star contracts. Most stars seem to change from red giants to pulsating variable stars, growing and shrinking, brightening and dimming, before they extinguish themselves for all eternity. Actually, there are several different types of these variable-brightness stars. Brunhilde, show table of variable stars."

Table 7.1
Variable Stars

Variable Stars	Oscillation Period	Star type	Distance Markers
Cepheid variables	1–70 days	Pulsating, large, luminous, yellow stars	More than 700 known in Milky Way; Distance markers out to 20 Mpc (66 million ly).
RR Lyrae variables	Less than a day	Pulsating blue-white giants	4,500 known in Milky Way; these stars are used to measure distances out to 200,000 pc (600,000 ly)
Mira variables	80–1,000 days	Pulsating red giants	1,000 known

pc = parsecs; Mpc = megaparsecs; ly = light-years

"Let's talk about my favorite variable stars, the Cepheids. In the early 1900s, Earth astronomers figured out that the longer the period of light oscillation, the greater the luminosity of the Cepheid star. This means you can determine the absolute magnitude of the Cepheids once you know the period."

Miss Muxdröözol puts her hand up. "The oscillation consists of a quick onset of brightness followed by a gradual dimming over and over again."

"That's right."

Mr. Plex begins to indicate the shape of the oscillation by jumping up and down like a lunatic released from an insane asylum:

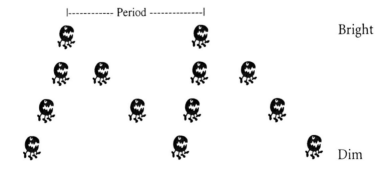

Mr. Plex illustrates Cepheid light output as a function of time.

"Miss Muxdröözol, could you retrieve my card with the distance modulus formula from your pocket. I gave it to you before when we were in your living quarters."

"Bob, here it is."

Bob, Distance Modulus Expert
Call me some time.

$$m - M = 5 \log\left(\frac{d}{10}\right)$$

She hands Bob the card. "I remember that d is the distance in parsecs."

"Right. Now, by looking at the period of oscillation of a Cepheid, we know the absolute magnitude M. Remember, this is because the period-luminosity relation for Cepheids gives us the absolute magnitude. By observing the star's brightness, we know the Cepheid's apparent magnitude m. Therefore we can calculate the distance from us to Cepheids. It's as if their blinking is literally calling out their distance from us. By measuring Cepheid variables across our cosmos, astronomers have been able to accurately measure vast distances between stars and between galaxies. Astronomers have used Cepheid variable stars to help them estimate the size of the entire Universe."

"Sir, is the Cepheid blinking better for determining distances than stellar parallax, which you already told us about."

"Damn right! Using space telescopes circling Earth, the Cepheids are wonderful distance markers out to about 20 Mpc (megaparsecs) or 66 million light-years. The first Cepheid variable was detected in 1784 by British astronomer John Goodricke. In 2000, astronomers showed that the Cepheid star Zeta Geminorum (in the constellation Gemini) swells and shrinks, making it the first Cepheid that astronomers have actually seen change its size. As Zeta Geminorum changes size, it flickers on a ten-day cycle."

Mr. Plex says, "During the pulsation, does the mass or luminosity of the star change?"

"No, only the outer envelope of the star appears to expand and contract."

Mr. Plex leans so close that Bob can feel Mr. Plex's cold breath on his shoulder. "Sir, are there many Cepheids?"

"When we left Earth, astronomers had discovered several thousand of these oscillating wonders in our local group of 30 galaxies that includes the Milky

Way and Andromeda. A typical Cepheid variable star is anywhere from 5 to 20 times more massive than the Sun, and it shines 100 to 10,000 times brighter than the Sun. A Cepheid variable star may oscillate for a million years, which, I remind you, is the blink of an eye relative to the several-billion-year life span of most stars."

"What about the other variable stars in Brunhilde's table?" asks Miss Muxdröözol.

"Those are other kinds of variable stars. RR Lyrae stars (named after variable star RR in the constellation Lyra) all have about the same luminosity (absolute magnitude), regardless of the length of their period. Since we know their apparent magnitude, we can calculate their distance from Earth using the distance modulus formula. The RR Lyrae stars are pulsating blue-white giants. Mira variable stars are cool but luminous stars typically vary in brightness by five to six magnitudes. The oscillation period is 80 to 1,000 days, so Mira stars are often called 'long-period variables.' Mira variable stars are complicated. When they were resolved by the Hubble Space Telescope, the stars did not appear to be spherical. They were irregular blobby shapes. One Mira variable, TX Camelopardalis, has been thoroughly studied. Astronomers discovered that when it contracts the star draws in nearby gas, and when it expands, it blows out an even larger amount. The star expels an Earth's mass of gas each year. Exactly how the star loses mass remains a mystery."

The ants look at Bob. "All your information is accurate." Ishmael points his golden abdomen at Bob. "You have not asked where we are from."

Bob feels stupid. The ants are right. That probably should have been among his first questions. "You'll have to excuse me. We've been through so much. Where are you from?"

Cain, the ant with the multicolored abdomen, speaks. "We have been following life on Earth since your Triassic led to the domination of dinosaurs. Our home world circled Betelgeuse, and, as the star grew, we survived by moving our planet into more distant orbits. We watched as mammals gradually arose on Earth and humans lifted their crafts to space. We eviscerated Einstein, Simon and Garfunkel, Brittney Spears, William Jefferson Clinton, and Shania Twain."

"What do you mean by eviscerated?" asks Bob.

"Perhaps I should say infiltrated. Several of your great scientists and artists were actually Betelgeuse ants holoprojected onto your planet."

"That's absurd," Bob says. "Why would you do such a thing?"

"We wished to shape you in subtle ways."

"For what purpose?"

"So that one day you would be ready to meet the Nephilim."

As the ants guide the chariot through the cemetery, the tombstones become more modern-looking. There is more order here. Bob imagines the tombstones are little molecules that suddenly crystallize in a supersaturated solution. After a few minutes, Bob notices fresh flowers placed on some of the newer headstones. It is as if Bob and his friends are riding forward in time.

Some Science Behind the Science Fiction

The sun, with all those planets revolving around it and dependent on it, can still ripen a bunch of grapes as if it had nothing else in the universe to do.

— Galileo

In this chapter, Bob recalled rumors that several Christian and Jewish sects had sprayed tiny message-carrying robots into the cosmos. These messengers were to orbit stars and await the possible awakening of civilizations on nearby planets. The general idea of messengers has been discussed in-depth by Ronald N. Bracewell, a leading radio astronomer from Stanford University.[6] According to Bracewell, advanced civilizations may prefer to use messengers instead of radio waves to make contact with other worlds. Certainly an extraterrestrial civilization could make itself more obvious to us, or make its signals more easily detectable, using messengers rather than radio signals that might be harder for us to find.

Consider a messenger sitting on our moon as it waits for civilization to evolve. It listens continuously for narrow-band emissions suggestive of a nascent civilization starting to use radio waves. Once the messenger detects the signal, it waits a century for Earth's science to mature, and then it simply sends back to Earth the detected radio signals, producing an echo that might excite our scientists. Perhaps variations in the echo times could actually contain a message from one of these machines.

Bracewell believed that the messengers might be "sprayed" toward nearby stars by advanced civilizations, and the messengers would not reveal themselves. They would merely report to the aliens when they heard signs of intelligent life, using a star-to-star relay system for communication. If a messenger were listening to our TV signals today, what would they be transmitting from us to their alien progenitors—"The X-Files," "Oprah Winfrey," "Seinfeld," "The Simpsons," and WWF wrestling matches? Would the alien civilization, someday, years in the future, be pondering the utterances of Howard Stern or Rush Limbaugh?

Many have speculated on what the messengers might look like.[7] There is no reason for a messenger to have a head, body, arms, and legs, even if the aliens had such appendages. Judging from our own experiences with robots, limbs would be too fragile and breakable and could malfunction. More likely, the messenger would have a compact, sturdy shape such as a sphere or icosahedron. If the messengers required projections for sensing, receiving, or mobilizing, the projections would be redundant or self-repairing to avoid mishap (figure 7.7).

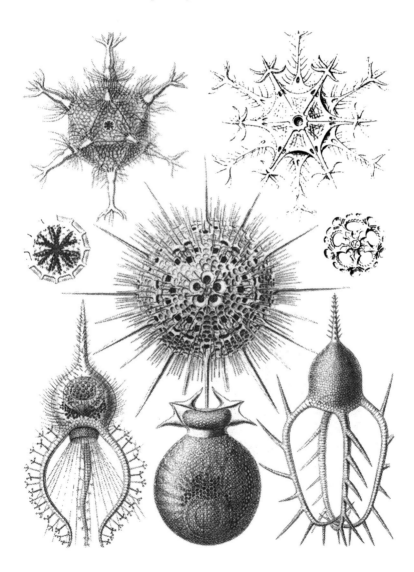

Figure 7.7   Messengers from outer space. [From Ernst Haeckel, *Art Forms in Nature* (New York: Dover, 1974).]

Another possibility is that the alien civilization intends for us to meet messengers resembling the aliens to assess how we react to them. In the scenario where aliens are aware of our own appearances, they might construct a messenger to resemble us so that we would feel more comfortable interacting with it.

I like to imagine the faint possibility that supercivilizations are already linked in a galactic federation of intelligent beings. Perhaps they are experienced in making contacts with emerging intelligence such as our own. If there are superintelligent, technological races in our Galaxy, then the messengers may already be here in our Solar System, hibernating in wait mode. This is a safe way for the aliens to gain or give information without making the dangerous interstellar voyage. There may be thousands of messengers swarming in the asteroid belt, reproducing using the large deposits of metals in this region. Their antennas might be pointed at Earth right now—waiting for the next Einstein, Jesus, or Mother Teresa to hit our air waves . . . Perhaps by monitoring major telephone microwave links between New York and New Jersey or various communication satellites, aliens could be scanning and downloading the entire contents of the Internet's world wide web as they searched it for works of art, music, science, and literature. Whether they like it or not, they would also be downloading the ever-increasing pornography, romantic discussions, moneymaking schemes, Pamela Anderson photos, conspiracy theories, eBay auctions, and all manner of the wild and weird.

* * *

Let's discuss further the different types of stars by starting with an analogy. Developmental biologists staring at embryos floating in bottles of formalin preservative can visually classify the specimens by age groups. In their minds, biologists carry an image of the entire developmental cycle of the organism under study. For example, in the third week of human development, the embryo exhibits a closed tube in which the brain and spinal cord are to develop. At the beginning of its fourth week, the embryo, now about 4 to 5 mm (about 0.16 to 0.2 in.) long, has a tail and the beginnings of eyes and ears, and the neck has gill clefts. Early in the second month, the buds of the arms and legs appear. Astronomers are in a similar position with respect to the evolution of stars. By classifying stars in the various stages of their development, astronomers can understand the entire process of stellar birth and death. We've discussed how when the Sun dies, it will become a red giant and fuse helium into carbon and oxygen. Betelgeuse in the constellation Orion, and Aldebaran in constellation Taurus, are already in this stage of evolution. If you look up in the night sky, you can see their red coloration. When the Sun becomes a red giant, its volume

increases and it surface temperature decreases, but this decrease is not sufficient to protect the Earth from ruin. One of the first things we will notice as
our Sun dies is that the Earth's polar ice caps will melt. The excess liquid water
will sink our coastal cities, and thick layers of clouds will temporarily hide the
Sun. For a time, life might flourish. A hot and humid atmosphere will envelop
the world, and vegetation will thrive everywhere.[8]

This fecund growth almost sounds nice, a return to our primeval African
roots. Perhaps when it gets too hot for the plants, we can burrow under ground
and try to live off the sulfur-based ecosystems of deep-ocean vents. We'll be
dining on huge tube worms that thrive on the ocean floor. There's just one big
problem. Soon the Earth's atmosphere begins to evaporate into space. Because
there is no atmosphere, the sky is black. The Sun is a huge, red orb that covers
half the sky. Daylight is 3,000 times more intense than it is now. The skies become clear, and the luxuriant vegetation catches fire. Soon the Earth looks like
the moon. Nothing lives. The oceans are gone. Wait a few hundreds of thousands of years, and the rock begins to melt, and lava flows across the face of the
land and pours into the empty basins that once held the blue oceans.

Depending on the rate at which the Sun loses mass on its way to becoming a
white dwarf, there are different fates for the Earth. If mass is lost early on, the
dry Earth could escape being vaporized as its orbit expands in response to the
Sun's lower mass. As a result, the Earth would orbit near Mars's current position as a cenotaph to humankind's past glory. It might be a featureless cenotaph, and I wonder if we could trace some of Earth's ancient shorelines and
ocean basins. On the other hand, if the mass loss occurs late in the Sun's evolution, the Earth may actually orbit inside the outer solar atmospheres. In this
case, the Earth would be burnt to a crisp and pulled deeper into the Sun.

There is also the possibility that another star could pass Earth before the Sun
became a red giant. Although this is very unlikely, a red dwarf could wander by
and toss Earth out of the solar system. Due to the gradually diminishing light,
the Earth would cool and the plants would freeze. However, inside the Earth or
at the bottom of oceans, life might have no trouble persisting for billions of
years. If the Earth actually got trapped into orbiting the red dwarf, life might
thrive for a very long time.

Astronomer Bruce Elmegreen predicts that the Sun will engulf the inner planets,
like Mercury and Venus, and vaporize them because the Sun's surface temperature
would be approximately 4,000 degrees Kelvin. The Sun would also be brighter so
that the temperature on all the non-engulfed planets would increase by five to six
times. This temperature on Earth would certainly vaporize the water and sulfur
and drive out most of the gases from the atmosphere. A slightly higher temperature
would vaporize silicon rocks. Luckily for any Earthlings who fled to the Solar System's

outer planets, these planets would not be vaporized, but whatever tenuous atmospheres they had would certainly change. Perhaps the outer planets would have spring-like temperatures, and Mars would likely be spared.

At some point, the Sun becomes so extended that the surface gravity is too weak to hold the atmospheric gas dust, and this mixture blows out of the star as a stellar wind. The Sun in this state is called a planetary nebula, with a hot central blue-violet star surrounded by rings of yellow and red. The rings are made of material from the Sun that has been blown into space. The core becomes a white dwarf (figure 7.8).

Planetary nebulae display an amazing variety of structures and sizes, and their incredible beauty makes them my favorite objects in the Galaxy. Some are round; some display structures with two lobes; some are so strange as to defy description. Many display several shells of gas.

When the Earth and Sun die, some of the elements in our buildings and bones are returned to the galactic graveyard to form new stars in a new cycle of

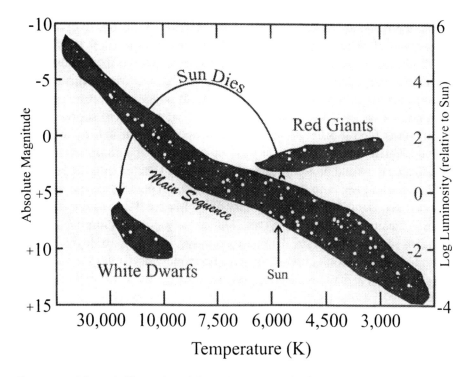

Figure 7.8 Schematic illustration of the evolutionary path of a star of one solar mass on the H-R diagram. When the Sun has used up all of the fuel in its core, it leaves the Main Sequence to become a red giant. Finally, the Sun becomes a planetary nebula, blowing off its outer layers. The remaining dead star is a white dwarf with a carbon-oxygen core.

birth and death. How might we save ourselves from these interminable cycles of birth and death? We might buy ourselves more time by migrating to some of Jupiter's water-containing moons. We could also use numerous powerful rockets to move the orbit of Earth so that it is further away from the Sun. Or we could use the stellar rejuvenation process, discussed in chapter 2, that employs hydrogen bombs launched into the Sun. All these methods may buy us more time so that we could safely leave the solar system entirely and achieve transcendence in nearby stars or in huge arks floating in space.

* * *

In this chapter we also discussed the Cepheid variables, stars whose periods (the time for one cycle of dimness and brightness) are proportional to the stars' luminosity. Using the distance modulus formula, this luminosity can be used to estimate interstellar and intergalactic distances. The American astronomer Henrietta Leavitt (1868–1921) discovered the relationship between period and luminosity in Cepheid variables (figure 7.9). I think of her on par with the

Figure 7.9 Henrietta Swan Leavitt (1868–1921). (Drawing by K. Llewellyn Blakeslee.)

early stellar parallax observers, because, in a sense, she was the first to discover how to calculate the distance from the Earth to galaxies beyond the Milky Way.[9] In 1902 she became a permanent staff member of the Harvard College Observatory and spent her time studying photographic plates of variable stars in the Magellanic Clouds. In 1904, using a time-consuming process called superposition, she discovered hundreds of variables in the Magellanic Clouds. These discoveries led Charles Young of Princeton to write to Harvard College Observatory director E. C. Pickering, "What a variable-star 'fiend' Miss Leavitt is; one can't keep up with the roll of the new discoveries."[10]

Leavitt's greatest discovery occurred when she determined the actual periods of 25 Cepheid variables, and in 1912 she announced the famous period-luminosity relation: "A straight line can be readily drawn among each of the two series of points corresponding to maxima and minima, thus showing that there is a simple relation between the brightness of the variable and their periods." Leavitt also realized that "since the variables are probably nearly the same distance from the Earth, their periods are apparently associated with their actual emission of light, as determined by their mass, density, and surface brightness" Sadly, she died young of cancer before her work was complete. Because of her astounding discoveries, in 1925 she was nominated posthumously for the Nobel prize.

* * *

In our tale, Mr. Plex was fascinated to learn that in low-mass stars the onset of helium fusion can be very rapid, producing a burst of energy called the helium flash. We expect a helium flash to occur in stars like the Sun because the core is in a "degenerate" state. This means that the core has contracted so much that the pressure of electrons in the core prevents it from contracting further. Under normal gas conditions (the usual nondegenerate state), an increase in the temperature of the core causes an increase in core pressure. This pressure causes the core to expand and the temperature to drop. This balanced state occurs when the star is in hydrostatic equilibrium. With a degenerate core, the temperature increases but the pressure doesn't. This extra energy ignites the helium, creating runaway nuclear reactions. This ignition is referred to as a "helium flash."[11] The helium flash lifts the solar core out of degeneracy, and the Sun enters a relatively stable phase in which helium fuses into carbon.

To better understand this process, consider that by the time a star like the Sun reaches the red-giant region in figure 7.8, the helium-rich core has been compressed to a volume about twice that of the Earth. At this extreme density, the gas behaves more like a solid than a gas. The core does not expand with an increase in temperature. Because the core cannot expand, it cannot cool. The rate of helium burning increases with temperature. More energy is generated, which further increases the temperature. In a few hours, the temperatures leap to hundreds of millions of degrees, causing the generation of as much energy as 100 billion stars of solar luminosity, or as much energy as the whole Galaxy generates. This result is the explosive helium. Instead of tearing the star to shreds, the helium flash's energy goes into removing electron degeneracy, and this alters the star so that it may now burn helium in the core, which acts as a perfect gas. We've never seen a helium flash. Maybe we never will. The flash occurs deep within the star, hidden from us by vast quantities of gas. The helium flash is generated in computer models of stars, and most astronomers believe it happens, though we will never actually observe it.[12]

Bob's description of the energy-producing fusion of hydrogen to produce helium is correct, but a simplification of the actual process. The *proton-proton reaction* is actually a chain of thermonuclear reactions that are the main energy sources for the Sun and cool, Main Sequence stars.[13] In the first step of the reaction, two hydrogen nuclei ^{1}H (shown as two protons, each symbolized by "+" in the rectangle of figure 7.10) combine to form deuterium ^{2}H. One of the

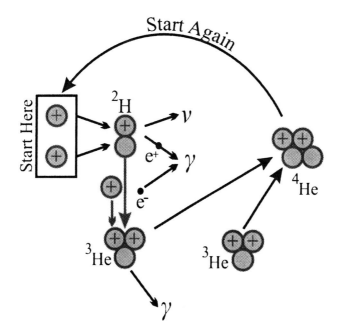

Figure 7.10   The proton-proton reaction starts with two protons (hydrogen nuclei in the box) and eventually produces a normal helium atom, ^{4}He, and releases energy. This is the main energy source for the Sun.

protons instantly turns into a neutron and ejects the positive charge as a positron e^+. This process is also accompanied by release of some energy in the form of a neutrino v. (In 1988, researchers in a Japanese metal mine performed an experiment that detected neutrinos coming from the direction of the Sun. However, there were only about half as many as expected. American researchers first discovered the solar neutrino problem in the 1970s.)[14] As soon as the positron collides with a free electron, the two destroy one another in a burst of energy in the form of a gamma ray γ.

In the next step, the deuterium combines instantly with another proton (+) to create a light form of helium called ^{3}He. Finally, after about a million years have passed, two ^{3}He nuclei combine to form regular helium ^{4}He with the ejection of two protons (and the process starts again). Notice that the net result is one helium atom and the release of energy.[15] This is the reaction that keeps us warm, happy, and alive.

Although you might like to think of the Sun as a large hydrogen bomb, it differs in two important respects from a nuclear weapon made on Earth. For one thing, certain reactions are very slow. For example, it would take each proton in the core of the Sun ten billion years to find and react with another proton to make deuterium (the first step in figure 7.10). (Subsequent reactions in the proton-proton chain are much faster.) The reason this first step works at all is because there are so many protons in the Sun. However, we should be thankful that the first step is so sluggish because this allows the Sun to consume its fuel very slowly over billions of years. Marcus Chown, author of *The Magic Furnace*, notes that the reaction between two protons to make a nucleus of deuterium is so slow that even the human body generates more heat, volume for volume, than the Sun! The Sun is so hot, despite such an amazingly low rate of heat production because it is huge. Like all large bodies, its surface area is small compared to its volume. This means that solar heat, which can escape only through its surface, mostly gets trapped inside.

* * *

Bob explained how by measuring Cepheid variables across our cosmos, astronomers have been able to accurately measure large distances between stars and between galaxies. Today, astronomers also search for exploding sun-like stars called Type 1a supernovae to estimate distances. Within a week of exploding, all Type 1a supernovae reach the same peak luminosity—about equal to 100 billion stars in a typical galaxy. Because the observed brightness of a supernova falls in proportion to its distance from Earth, astronomers can calculate the distance to each supernova. (These kinds of supernova are difficult to find because each galaxy only produces about two of them every thousand years.) The current information on supernovae distances seems to suggest that the expansion of the Universe is speeding up.[16]

I thought of a labyrinth of labyrinths, of one sinuous spreading labyrinth that would encompass the past and the future and in some way involve the stars."

— Jorge Borges

The thing's hollow—it goes on forever—and—oh my God!—it's full of stars.

— Arthur C. Clarke,
2001: A Space Odyssey

CHAPTER 8

Stellar Graveyards, Nucleosynthesis, and Why We Exist

And there we saw the Nephilim, the sons of Anak, who come of the Nephilim; and we were in our own sight as grasshoppers, and so we were in their sight.

— Numbers 13:33

"I'm a hybrid," Miss Muxdröözol says. "That's what you wanted to know."

Bob looks out the window and then back at her. "Well, I was curious."

"My great grandfather was semi-aquatic and spent his life among telephone cables that ran across the bottom of the Gulf of Finland, just outside Leningrad harbor."

"This is bizarre. No need to tell me more."

"You said you were curious. Just let me finish. He was a spy, eavesdropping on Russian military communications between the Russian naval base at Sosnovyy Bor and the military headquarters at Zelenogorsk. Apparently he could confuse the communications with his dreams, flipping bits every now and then, thereby ending any remnant of the Cold War."

"But you're mostly of trochophore lineage?"

"Yes, I closely resemble my mother."

Miss Muxdröözol plays nervously with a few of her thoracic vertebrae. Apparently she can remove and replace them at will with no deleterious effect to her well-being or body's integrity. Perhaps she can regenerate them at will and need not return the portable vertebrae to her spinal column.

Ishmael puts his forelimb on Bob's leg. "Bob, could you finish your discussion of stars? We're almost ready to see the Nephilim."

"And believe me," Cain says, "that's an experience you'll never forget."

Bob nods. "All stars die, Miss Muxdröözol."

"Do they evolve in the same way, like you've been telling us?"

"Yes, very similar ways although at different rates. Finally their cores become mostly carbon. Brunhilde, display aging star with carbon core." Brunhilde redisplays figure 7.6.

"I love carbon, Sir. My body is made of diamond, the hardest naturally occurring known substance in the Universe."

"Yes, Mr. Plex, carbon is special. It's everywhere in nature, and it forms more compounds than all the other elements combined. We all owe our lives to the carbon core of aging stars."

"Sir, why do so many different kinds of carbon compounds exist?"

"The reason is that carbon can form so many different kinds of chains, ring structures, and three-dimensional arrangements. Carbon is in our proteins and our genetic material. It's in coal, petroleum, and natural gas. It's in the atmosphere, in steel, in dyes."

Brunhilde speaks of her own volition. "Until about A.D. 300 on Earth, ink was made of carbon mixed with gum and water."

Bob smiles. "Brunhilde, you are a wonderful repository of facts. Thank you." Bob strokes the piezoelectric undersurface of the flexscreen, and Brunhilde makes a soft purring sound. "But let's return our discussion to the death of stars. The exact final state of a star depends on its mass. A star less than 1.4 times the Sun's mass dies calmly. Very massive stars explode before their death. Stars about the mass or our Sun become fat red giants when all the helium fuel is consumed."

Miss Muxdröözol starts playing with some of the flexscreen's knobs and then puts the device face down on the chariot's seat. Evidently Miss Muxdröözol is curious about Brunhilde and the precise nature of her relationship with Bob. "What does the dying star do next?" Miss Muxdröözol asks. "What happens exactly?"

"The star shoots some of its mass into outer space. In particular, the star's outermost hydrogen envelope, which by now has lots of heavier elements, blasts into space. Electrically charged particles flow outward as a stellar wind. Deep layers are also tossed off in a tenuous shell of gas called a planetary nebula. We talked about them before. I'm reviewing. Only the star's core remains—like a naked peach pit after the peach has been eaten away."

"Or a naked pearl," Miss Muxdröözol says, "after the clam shell has drifted away into the vast ocean."

"Very poetic," Bob says. "Earth's astronomers have catalogued over 1,600 *planetary nebulas*, but we know there were tens of thousands in the Milky Way. They were probably less than 50,000 years old, because the nebula quickly diffuses away."

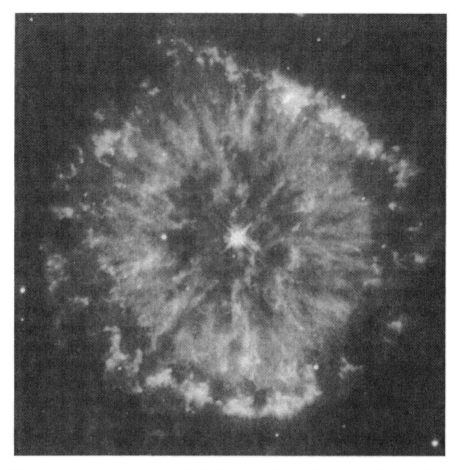

Figure 8.1   Astronomers using NASA's Hubble Space Telescope have obtained images of the
planetary nebula NGC 6751. Glowing in the constellation Aquila like a giant eye, the nebula
is a cloud of gas ejected several thousand years ago from the hot star visible in its center.
[Figure courtesy of NASA and The Hubble Heritage Team (STScI/AURA).]

Brunhilde speaks. "Bob, if you turn me over, you will see that I am now
displaying the beautiful planetary nebula NGC 6751" (figure 8.1).

"Sir, why do they call them 'planetary' nebulas? There's nothing planetary
about them."

"Well, as I hinted at the last time we spoke of these objects, they are bright
nebulae that sort of resemble planets when viewed through a small telescope.
But you are right, they are really expanding shells of bright gas. The term nebulae or nebulas refers to various kinds of gas and dust clouds in interstellar space.
Most known planetary nebulae have an extremely hot central star with a temperature of up to 150,000 degrees Kelvin."

Bob looks out the window. The land around the road resembles an Oriental carpet—lavender, crimson, and orange wildflowers mingled with silvery green prairie grass. Nearby, a crystal stream rushes over a white bed of pebbles. It is hard to tell if the stream is natural or artificial.

"Now, close your eyes. I want you to imagine the remnant of the Sun, a carbon core surrounded by a shell of burning helium. The nuclear fuel is exhausted. The pull of gravity is immense and there's nothing to counteract it. The star contracts, the temperature and pressure grow, and electrons are stripped off atoms. The star is now a small, hot white dwarf. Brunhilde, show white dwarf."

Upon the screen appears a small, hairy, white creature with a huge grin:

White Dwarf

"Bunhilde, that is not what I meant!"

"Sir, I believe Brunhilde is beginning to show signs of malfunction," Mr. Plex says.

"Brunhilde show stellar white dwarf in comparison to Earth's size."

Bob looks at the odd image. "Well, Brunhilde is in a playful mode. She got it partially right. Let me repeat some facts I told you a few days ago. Eventually the white dwarf star has a diameter roughly equal to the Earth's diameter. It is made of electrons and nuclei that are squeezed more closely than normal atoms can pack. At this point, they can't pack any tighter. The gravity on the star is about 350,000 times greater than on Earth, which would mean that you, Miss Muxdröözol, would weight 350,000 times more than you do on Earth. A teaspoonful of a white dwarf, if brought to Earth, would weigh as much as an elephant."

Brunhilde has apparently overheard Bob's conversation about the weight of elephants and white dwarfs because the flexscreen is showing an amazing but chaotic array of symbols

Mr. Plex looks at the screen. "Sir, it's time to reboot her."
"I suppose you are right. Brunhilde, reboot Windows 2100."
The screen displays:

$$\boxed{\text{𝕾𝔥𝔲𝔱𝔡𝔬𝔴𝔫, 𝕷𝔬𝔤𝔬𝔣𝔣, 𝕮𝔥𝔞𝔫𝔤𝔢 𝕻𝔞𝔰𝔰𝔴𝔬𝔯𝔡, 𝕽𝔢𝔰𝔱𝔞𝔯𝔱?}}$$

Bob selects "Restart." Within two seconds Brunhilde is rebooted, and Bob
says, "Display Sun's life stages." Brunhilde is apparently working just fine and
displays the life cycle of sun-like stars (figure 8.2).

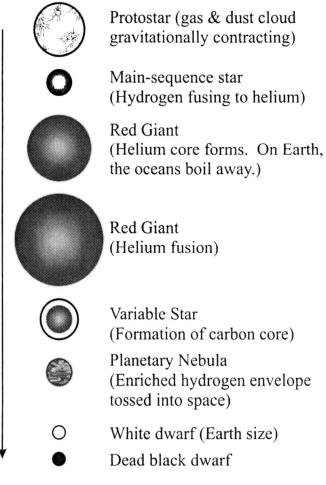

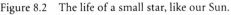

Figure 8.2   The life of a small star, like our Sun.

"Sometimes a white dwarf in its final death throes becomes a nova, a bright flaring star. As far as we know, all novae are white dwarfs with a companion star, and portions of this other star may fall into the white dwarf and fuel the quick flaring. As the white dwarf cools, it turns a dim red, and finally becomes a dead *black dwarf* in the interstellar graveyard.[1] How long the transition to black dwarfs takes is uncertain. It could be on the order of trillions of years. The black dwarf is the size of Earth but with a density 50,000 times that of water, covered possibly with a thin layer of ice and surrounded by an atmosphere a few meters thick." Bob pauses. "There are lots of graveyard states for stars, and which of these graveyards the stars evolve into depends largely on the star's mass. White and black dwarfs are just some examples. Later we'll talk about other graveyards, like black holes, neutron stars, and pulsars."

Bob looks at the ants. They are so still. If Bob and his friends are being taken to the Nephilim—a race of people mentioned in the Bible—does this mean that the ants believe in God? "Excuse me," Bob says to Ishmael, "does your race believe in religion and God?"

The ant shoves the electrical probe up its nostril. "We have followed your race for a long time. Your people developed self-aware machines, like Brunhilde, in 2021. Intelligence is an emergent property—it appears spontaneously in systems of sufficient complexity. Several years after the Valkyrie membrane abducted you, your Internet itself became a noncorporeal intelligence that arose through random fluctuations in a network of wires and electromagnetic signals. Your 'Internet' stretched across planets, stretched to other stars. Even people in the smallest villages on Earth were downloading information from Alpha Centauri via a wormgate system that linked wormholes. The wormholes folded space and appeared to permit signals to travel faster than light by stepping around ordinary space."

"Where are you going with this story?" Bob says.

"One day, a villager in Ethiopia entered a strange URL into his web browser: http://www.god.ac, the suffix 'ac' represents domain names on planets in the vicinity of Alpha Centauri. At this point, your species discovered God, and we discovered him a few days later."

Bob thought about this. Were the ants telling him the truth or was this some kind of elaborate joke or puzzle for him to solve? What would God's web page be like?

The ant continued. "Of course the time for interstellar downloads normally would be quite slow. But your wormhole modems took care of that. We all discovered that God existed and that he was a shrimp-like crustacean living beneath the soil of the planet Mercury."

Bob gasps. "That's crazy." Surely the ants must be mistaken. Humans were supposed to be made in God's image. Of course, the word 'image' may have referred to human mental apparatuses rather than physical appearance. And now that so many different extraterrestrials have been discovered, perhaps it was true that God could take many different forms.

Bob shakes his head. "I'm almost done with our star lessons. Brunhilde, redisplay diagram showing the fate of sun-like stars." Brunhilde redisplays figure 7.8.

"This schematic diagram tells much of the story, but it is simplified. Let's have a final, complete review. Sun-like stars start their life as a protostar, which gravitationally contracts, until the star is located somewhere on the Main Sequence. Next, the 'stable' Main Sequence star shines by fusing hydrogen to helium. The star evolves to a red giant when it exhausts the hydrogen in its core, and the hydrogen ignites in a shell around the exhausted core. Later, helium fusion occurs when ash from the shell builds up in the core causing it to contract and heat up. Finally we have a variable star with a carbon core, followed by a planetary nebula with its hydrogen envelope and heavy elements tossed into space. At the bottom of this figure we have the final white dwarf. The white dwarf eventually dies when it turns into a black dwarf, which would be located further down and to the right on this diagram."

The chariot stops. The door opens. Ishmael touches Bob with his forelimb. "The Nephilim are here," Ishmael says.

Slowly, a white spherical opening appears. It floats in the air about three feet above the violet grass. The world around Bob dims. A hush falls over the landscape; even the rodents on the ground are motionless and stare up in apparent wonder.

Miss Muxdröözol grabs Bob's hand. Mr. Plex and Brunhilde gasp in unison. The ants smile, their strange mandibles opening wider and wider. A bird falls to the ground. The odor of limes is everywhere.

The Nephilim have some characteristics that Bob is familiar with. They are winged like angels and have long trailing beards as Bob imagined the biblical prophets to have. But that is where the similarity ended. Their arms are covered with bubbly skin that occasionally pulsates like little balloons inflating and deflating. Their eyes are iridescent balls, the size of avocados, with hard crystalline corneas.

Bob looks closer at their faces. Their eyes are moist, and each appears to have three separate pupils. Their hands are insectile. Each creature has six wings. With two of the wings they occasionally cover their faces.

In a whisper, Mr. Plex begins to recite from the Bible. "It says in Numbers 13:33, 'And there we saw the Nephilim, the sons of Anak, who come of the Nephilim; and we were in our own sight as grasshoppers, and so we were in their sight.'"

Bob turns to the ants. "Do they speak?"

"They are very difficult to understand. Our brains are not wired to fully appreciate their ways of thinking. And they certainly have some odd interests and collections."

The temperature drops noticeably. The stars in the sky seem to wink out, but perhaps it is just an illusion. "Come this way," the Nephilim say. Their sharp, thin "claws" retract with a whir. Their folded wings, when viewed from different angles, seem to form pictures or shapes, the content of which depends on who is viewing the wings.

Bob grabs Brunhilde. Miss Muxdröözol holds Bob's other hand as they walk through the grass. After several minutes they approach a great marble cube, on which are inscribed the cryptic symbols:

$$\downarrow$$

בְּרֵאשִׁית בָּרָא אֱלֹהִים אֵת הַשָּׁמַיִם וְאֵת הָאָרֶץ   וְהָאָרֶץ

Just inside the front door Bob passes through a metal detector operated by two winged angels. One of the angels wears a beige Armani suit, the pants are held up by suspenders. The other angel is huge, with long white hair. His several-hundred-pound body resembles the Great Pyramid of Giza. Christmas lights are hanging from the ebony walls. Bookshelves are filled with various Bibles: the King James English, Luther German, Louis Segond French, Latin Vulgate, New International, New American Standard, Revised Standard, Darby Translation, and many others, the origin of which Bob cannot decipher.

"Come to our nursery," say the Nephilim.

Bob, his friends, and the ants walk through the "nursery." There are lots of sounds. An electrical hum. The sloshing of water or other liquids. The ants' clattering legs against a hard floor. The Nephilim come to an abrupt stop. "Our pride and joy is our *Pikaia* collection."

Ishmael interrupts with a whisper, "*Pikaia gracilens* to be precise." The ant pronounces it pih-KAY-ah GRASS-ih-lenz.

Bob peers into a huge tank of small swimming creatures. Each animal is about 1½ inches in length. They swim above the gravel using their bodies and expanded tail fins.

Stepping close, Miss Muxdröözol asks, "What's so special about them?"

Bob replies, "I've heard of them. *Pikaia* is a representative member of the chordate group from which humans probably evolved. But these things died out before humans walked the Earth."

The Nephilim nod. Their wing feathers move forward in what Bob now recognizes as a Nephilim sign of affirmation. "We took specimens very long ago."

The ants nod. Ishmael's abdomen throbs as he starts to speak. "Humans discovered fossil Pikaia creatures in the Burgess Shale of Canada. They lived around 530 million years before humans. These kinds of creatures were ancestors of the chordates and precursors to vertebrates like yourself."

Brunhilde speaks. "Chordates have a stiff, dorsal supporting rod called the notochord. It organizes the nervous system. In later vertebrate development, the notochord becomes part of the vertebral column."

They walk through the Nephilim's zoo, and Bob gazes into the next tank. There is a single swimming creature with powerful, V-shaped longitudinal muscles rippling on each side of its long body. Because the water is murky, Bob cannot see the end of its body, only its tentacled head and a portion of its body. Having studied biology in college, Bob recognizes the floating creature in an instant. It is an amphioxus, also known as a lancelet.*

A fishy odor begins to fill the Nephilim's nursery, quickly followed by the smell of absinthe. In the distance, Bob notices an octopus-like robot dangling several jointed arms into the tanks of different creatures. Perhaps the robot assists with cleaning or feeding.

What is the amphioxus perceiving about Bob?

Billions of years ago, the genetic code of primitive cells was passed from generation to generation and finally to multicellular organisms, then to invertebrates, and then to vertebrates around 600 million years ago. Because amphioxus, Pikaia, and their kin are transitional forms between invertebrates and vertebrates, the amphioxus can be thought of as the evolutionary gateway to humanity, to more developed brains, and to consciousness. Amphioxus, the gate, a creature born millions of year ago... Why do humans exist? The answer quite simply is because amphioxus and its kin were born eons ego and lived to survive.

---

*"Lancelet" is the common name for about 25 species of simple marine animals, which are classified between invertebrates (animals without backbones) and vertebrates (animals, like humans, with backbones). They have a stiff dorsal rod and a notochord, but no vertebrae or heart. Around their mouths, cirri and tentacles move like a bag of nervous worms. They share the same ancestry as amphioxus, as evidenced by their embryonic tongue-barred gill slits.

Bob passes other exhibits in the distance. In one huge pool swim icthyosaurs. They resemble dolphins but are actually aquatic reptiles that swam in the Jurassic Period and became extinct on Earth long before humans evolved.

They finally arrive at a large rotunda that has an elegant fractal design laid into its marble floor. Large stained-glass windows are on all sides. Various Nephilim are milling about, often speaking in unison, as if they are part of some hive mind.

"Bob," says Ishmael, "Please finish your discussion of stars."

"Here? Now?"

"Yes, the Nephilim will it."

"Sir, please continue," says Mr. Plex. "I want to complete my education."

Bob smiles. "Okay, we just have a little left to cover. The most important information actually." Bob turns to his friends, "Humans, scolexes, Betelgeuse ants, Miss Muxdröözol, and even Brunhilde's organic brain are made of heavy elements like carbon, oxygen, potassium, nitrogen, and iron."

The Nephilim speak: "When the Universe was born, there was only hydrogen and helium. But, as you know, stars fused hydrogen into heavier elements. These elements are in your bodies." They gesture to a painting of Carl Sagan hanging on a rotunda wall. "As he always liked to say, 'We are all star-stuff.'" The Sagan on the wall starts to grin and then returns to its static form.

Bob nods, "The real miracle of life is the miracle of helium burning. Normally the nucleons in helium atoms are very tightly bound. To produce carbon, two helium nuclei must bind for sufficiently long until they are struck by a third helium nucleus—three helium nuclei provide six neutrons and six protons, the recipe for carbon.[2] The first step looks like this." Bob takes the lipstick from Miss Muxdröözol's pocket and begins to draw on the rotunda wall:

$$^4He + {}^4He \Rightarrow {}^8Be,$$

"However, the lifetime of beryllium-8 is only 10^{-16} seconds, a duration that would normally be insufficient to allow beryllium-8 to have time to go to carbon, the next step:

$$^4He + {}^3Be \Rightarrow {}^{12}C.$$

Miss Muxdröözol puts her hand up. "Wait, what do all the little numbers mean by 'He' and 'Be'?"

"Chemical elements, like hydrogen and helium, are determined solely by their numbers of nuclear protons. This quantity is called the *atomic number*. For example, hydrogen (H) has one proton, helium (He) has 2, carbon (C) has 6, and nitrogen (N) has 7. The sum of the protons and neutrons in the nucleus is that atom's *atomic mass*, written as a superscript before the chemical symbol. Ordinary helium, helium-4, has two neutrons and two protons. Ordinary carbon is carbon-12, but there are also carbon-13 and carbon-14 isotopes, which have additional neutrons. Ordinary beryllium is beryllium-9. The first formula might be symbolized by a diagram in which black circles are neutrons and unfilled circles are protons." Bob draws the diagram:

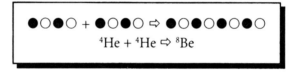

$$^4He + {}^4He \Rightarrow {}^8Be$$

"Okay," Miss Muxdröözol says, "So the beryllium-8 produced by the collision of two helium-4 molecules does not exist long. So, how is there time for the beryllium to combine with helium-4 to produce carbon?"

"We need three helium-4 nuclei to come together practically simultaneously to create the carbon."

Ishmael raises his golden forelimb, "This is the miracle of the Universe, and we feel that because the reaction is so difficult, it is proof of God's existence." He takes the lipstick from Bob and draws on the wall:

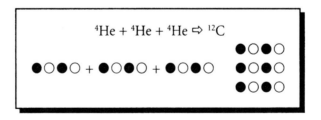

$$^4He + {}^4He + {}^4He \Rightarrow {}^{12}C$$

Bob nods. "This is sometimes called the miracle of *alpha capture*, because the helium-4 nucleus is also known as an alpha particle. The ant's formula is supposed to represent the final results of combining three helium-4 nuclei, but the formula does not indicate that it takes more than one step to carry this out."

Miss Muxdröözol is playing with her thoracic vertebrae when she looks at Bob and asks, "Why is helium-4 called an alpha particle?"

"Helium normally has two protons, two neutrons, and two electrons. An alpha particle is another name for the positively charged nucleus of the helium-4 atom. Alpha particles are emitted by some radioactive substances. But let me clarify why the triple-alpha process is so difficult and so wonderful."

"Triple-alpha process?" says Miss Muxdröözol.

"That's what astronomers called the combining of three helium nuclei to make carbon. Helium nuclei in the core of a red giant are moving about at high speeds. Encounters between the three nuclei, if they occurred, would be extremely brief. In the past, some scientists estimated that three nuclei would only have a thousand million million millionth of a second to react to form carbon."[3]

They pass some sculptures formed of twisted rocks and spires. They are crimson and others shades of lavender and turquoise, and they smell like cinnamon.

Bob sees several strange creatures polishing the spires. The creatures have the form of humans, but each has four faces and four wings. Their legs are straight, and their feet resemble calves feet, and they sparkle like burnished bronze. Under their wings are humanoid hands.

Bob continues. "But don't worry about this major difficulty—the briefness of simultaneous encounters between three helium nuclei to make carbon. As I mentioned, we can look at the collisions between *two* nuclei first to make beryllium-8. The average life span of beryllium-8 is a hundred thousand million millionth of a second. The good news is at least this is ten thousand times longer than the time that two helium nuclei spend together as they race past each other in the red giant's core."

"But the beryllium-8 dies fast!" Miss Muxdröözol says.

"Yes, but on atomic scales the beryllium-8 is lolling around, just waiting to be struck by another helium nucleus. So the two-step process to make carbon seemed quite possible to scientists in the early 1950s."

"Great," says Miss Muxdröözol.

They walk through new hallways with walls containing interlocking storage units. Each hexagonal opening is color coded; perhaps this indicates different food supplies or items necessary to maintain the huge zoo. The ceiling seems to bioluminesce, producing an eerie, orange light that washes over all of them.

"But there's some bad news. The speed of the two-step triple-alpha process to produce carbon depends on how *much* beryllium-8 resides in a star to be slammed into. And, we know that there is only one nucleus of beryllium-8 for every billion nuclei of helium. A very tiny amount! According to astronomers' initial computations, there was not enough time since the Big Bang for the triple-alpha process to have made all the carbon in the Universe. The nuclear reaction that jammed together beryllium-8 and helium-4 was just not fast enough to produce the large quantities of carbon in the Universe."

Bob passes a tank full of Moroccan trilobites displaying amazing genal, pleural, ocular, and occipital spinosity, a striking pygidium with raised radiating ridges and beautiful color variations. "Beautiful," he says as he stops and watches the creatures swim through the water. Trilobites were the most diverse group of extinct prehistoric animals preserved in the fossil record.

"Do you like them?" ask the Nephilim. "We recognize eight orders of trilobites, of which your scientists classified 16,000 species."

Bob nods as they pass tanks full of Paleozoic arthropods: Agnostida, Redlichiida Corynexochida, Lichida, Phacopida, Proetida, Asaphida, and Ptychopariida. Yes, Bob could spend a lifetime here looking at the remarkable panoply of primitive forms.

"Sir, don't keep us in suspense," Mr. Plex says. "You said the collision of beryllium-8 and helium-4 was not fast enough to produce the large amount of carbon in the Universe."

"This is what they thought at first. I'm giving you a bit of history here. The reaction of beryllium-8 and helium-4 seemed too slow. There was one chance that the reaction speed could be boosted—if carbon-12 had a very special property: an energy almost exactly equal to the combined energy of beryllium-8 and helium-4 at temperatures in a red giant. Chemists called this kind of facilitated nuclear reaction 'resonant.' If by some 'miracle' this were true, then the triple-alpha process could work."

A trilobite swims by and seems to shake its pygidium (tail piece) at Bob. Bob backs up into Mr. Plex and then continues his dissertation. "Here's some more history. First recall that the energy of an atomic nucleus has various energy states. The lowest energy state is called the ground state. Other states are called excited states. In the 1950s, astronomer Fred Hoyle realized that if carbon-12 possessed an excited state with an energy very close to the combined energy of beryllium-8 and helium-4, then the reactions could be speeded up—like inserting a key into a well-oiled lock that is carefully designed to fit the key. In particular, to overcome the 'beryllium barrier,' the excited state of carbon had to be quite close to 7.6 mega-electron volts (or million electron volts, MeV) above the ground state. If the beryllium barrier was overcome, then all the elements in nature could be made inside stars! Hoyle wondered if he was correct. The problem was that no one in the 1950s believed such a state of carbon existed. Nevertheless, Fred Hoyle in effect declared the state must exist because humans exist!"

Seven trilobites line up along the glass and stare at Bob with their holochroal eyes. Bob nods at them and continues. "And scientists finally discovered in the 1950s that a 7.6549 MeV energy state of carbon-12 did exist. This allows carbon to be created in stars. If the resonance level of carbon were only 4 percent lower, such intermediate bonding steps could not occur, and no carbon would be produced, making organic chemistry impossible.[4] And that, Mr. Plex, is one of the greatest wonders of the Universe." Bob hands Mr. Plex a card:

> The reason why life is possible:
> 7.6549

"Hold on," Miss Muxdröözol says. "Let me see if I get this right. In order for the carbon to be created in sufficient quantities for life, the carbon nuclei must have an excited energy level, which you call a resonance, very close to the sum of the energies of the beryllium and helium nuclei."

"You've got it! Here's one way to visualize a resonance. Brunhilde, display tromba marina (figure 8.3). Here you see an ancient instrument with a single string. The vibrating string is a metaphor for the resonance that exists in the excited nucleus of carbon-12. When the player presses the string down at different locations, different harmonics of the fundamental note sound. These higher harmonics are metaphors for higher energy states. The vibrations are constrained because the string is anchored at each end. Think of the carbon-12 resonance as a high note played by a musician on the carbon-12 string."

Nodding, she says, "It turns out that such a resonance exists in the carbon nucleus, and occasionally this excited state of the carbon nucleus is formed."

"Yes. Just like plucking a string creates a sound, the act of helium colliding with beryllium creates the excited form of carbon-12. Then the carbon-12 radiates energy away and settles into a ground state. Of course, reactions can go backward. The excited carbon could 'decay' back to three helium nuclei, but in approximately 1 out of 2,500 cases the excited carbon nucleus emits two photons and falls to its lowest energy state creating a stable carbon nucleus. It appears lucky that the resonance in the nucleus made from helium and beryllium nuclei is the appropriate value required to allow the formation of carbon."

The Nephilim nod. One actually appears to smile.

Bob nods back. "Here are the three main cosmic coincidences that allow us to exist. One: It's good that beryllium-8 exists for such a short time. If it lasted much longer, then nuclear reactions would rapidly convert all the star's helium

a

105

LXII   *Tromba Marina*

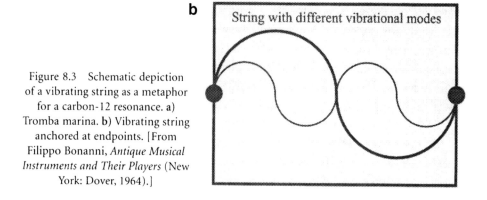

b   String with different vibrational modes

Figure 8.3   Schematic depiction of a vibrating string as a metaphor for a carbon-12 resonance. a) Tromba marina. b) Vibrating string anchored at endpoints. [From Filippo Bonanni, *Antique Musical Instruments and Their Players* (New York: Dover, 1964).]

to carbon, and this sudden release of energy would blow up the stars. There would be no heavier elements, no calcium for our bones, no iron in our blood."

More trilobites swim to the front of the tank and stare at Bob.

"Two: An excited carbon-12 state must exist very close to the combined energy of beryllium-8 and helium-4. In actuality the combined energy of beryllium-8 and helium-4 is 7.3667 MeV—a tad less than the 7.6549 MeV energy state for carbon-12."

"Sir, wait," Mr. Plex says. "Didn't you imply that the energy states had to be exactly equal?"

"It's okay. The red giant star is hot. The temperature is high enough to boost the energy of the He and Be nuclei to just above the critical threshold of 7.6549 MeV. The temperature has to be just right. The carbon excited state energy has to be just right. And they are just right."

"What's the third cosmic coincidence?"

"I call the third coincidence the *cosmic battle between carbon and oxygen*. It's a delicately poised battle where you don't want either of the 'warriors' to be too powerful."

"Explain." Miss Muxdröözol says.

"Okay, I've showed you how a fortuitous two-step triple-alpha process can create carbon-12. But what happens if another helium-4 slams into the carbon-12? No answer, Mr. Plex? Okay, I'll tell you. The carbon-12 is converted to oxygen-16."

"Isn't that good?" Miss Muxdröözol says. "We need oxygen? It's good that the sun can convert carbon to oxygen."

Hundreds of brightly colored trilobites inch their way up the tank's glass like plants in a fast-growing jungle. All around the rotunda, the Nephilim lean against walls as they listen to Bob, Mr. Plex, and Miss Muxdröözol.

"Yes, but all that work we did to create the carbon-12 is wasted if all the carbon quickly converts to oxygen. If that happened, the Universe wouldn't have enough carbon. We need *both* carbon and oxygen. All this means is that the reaction between carbon-12 and helium-4 cannot be too fast, but if it's too slow there wouldn't be enough oxygen to support life. Fred Hoyle was able to show that the fabric of the Universe had to be constructed so that the reaction between helium and beryllium is resonant, but, on the other hand, the reaction between carbon-12 and helium-4 is not resonant—in other words, the reaction is not fast, as would be the case if oxygen-16 had an excited state with an energy equal to the combined energy of a carbon-12 nucleus and helium-4 nucleus at red giant temperatures."

Miss Muxdröözol tosses one of her vertebrae to Bob. "And what *is* the excited state of oxygen-16?" she asks.

Bob holds the vertebra tightly in his hand. Perhaps this is a way that her species signals a desire for shared affection. "The combined energy of a nucleus of carbon-12 and of helium-4 is 7.1616 MeV. The excited state of oxygen-16 is 7.1187 MeV."

"Wow, that does sound too close for comfort. Doesn't the reaction that creates oxygen take off?" asks Miss Muxdröözol.

"No, notice that, luckily, the 7.1187 state is just *below* the 7.1616 MeV energy of carbon-12 plus helium-4. Thank God! The star's energy can *boost* the energy of reacting nuclei, but it can't decrease it. All this means that carbon-12 does not react too quickly to form oxygen. If the energy of oxygen-16 had been just a tiny bit above 7.1616 MeV instead of a tiny bit below, you and I would not be here." Bob stares into Miss Muxdröözol's large eyes and grips her vertebra even more tightly. "Life would not exist. All the carbon in stars would be converted rapidly to oxygen-16. No carbon, no creatures."

Miss Muxdröözol nods. "A fraction of a percent difference in energy changes the Universe."

"Yes. We exist because of cosmic coincidences, or more accurately, we exist because the seemingly 'finely tuned' numerical constants permit life. Some people who believe in the *anthropic principle* believe these numbers to be near miracles that possibly suggest an intelligent design to the Universe."[5]

Bob puts Miss Muxdröözol's vertebra in his pocket. Then he slaps his hands together to get everyone's attention. "We exist only because beryllium-8 is unusually long-lived for an unstable nucleus. We exist because carbon-12 has an appropriate energy state to permit its production. We exist because oxygen-16 does not have an energy state that would permit its rapid production and lead to the demise of carbon in our Universe. I do not know if God is a chemist, but the fabric of the Universe hinges on precarious nuclear reactions."

Bob is holding the flexscreen when Brunhilde suddenly says, "Bob, there's more, isn't there?"

"Yes, much more. In order for life as we know it to eventually arise, the star with triple-alpha capture must explode as a supernova."

"Of course," Brunhilde says.

"That's the only way to get a lot of carbon out of the star and into space. But supernovas can't be too common or else they would destroy too many worlds. On the other hand, as I mentioned before, supernovas are important because the shock waves they produce can cause planetary systems to start to coalesce from dust clouds surrounding other stars. This means that there can't be too many supernovas or too few. Any substantial deviation would decrease planetary formation and the emergence of life."

Bob taps on Brunhilde with Miss Muxdröözol's vertebra to get everyone's attention. "If the Sun were shrunk to the size of a sand grain, the closest star would be a few miles away. C. S. Lewis, the famous English writer, thought that God separated the stars by great distances so that one fallen species wouldn't spiritually infect another. But today we know that God may have had other reasons. He may have separated the stars to prevent supernovas from destroying too many worlds."

Brunhilde's voice takes on a low sensual tone as her piezoelectric surface oscillates to provide a fractal vibration to Bob's left palm. "Bob, tell us more about supernovas."

Frowning, Miss Muxdröözol says, "Brunhilde, please do not speak to Bob in that tone of voice."

"What's it to you?"

"He's mine."

Bob looks into Miss Muxdröözol's teardrop shaped eyes, which seem to be growing to the size of baseballs. "Miss Muxdröözol, I . . . I didn't know you cared."

Bob touches a button on the flexscreen and places Brunhilde in suspend mode. The illumination of her display dims by 50 percent. Bob continues. "Here's what I know about supernovas. Massive stars blow up when they die. In particular, stars about eight or more times the Sun's mass die as supernovas, huge stellar explosions" (figure 8.4).

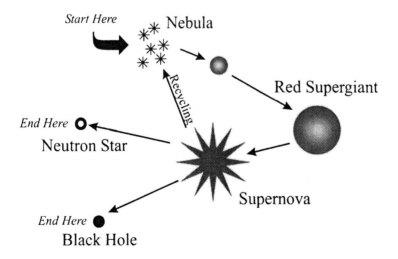

Figure 8.4   The life cycle of a massive star.
(See Jim Lochner, NASA, "X-ray Astronomy: Supernovae and their remnants,"
http://imagine.gsfc.nasa.gov/docs/science/know_ll/supernovae.html)

"Why is that?" says Mr. Plex.

"A massive star's carbon core contracts under the force of gravity just like a small star's does. But there's one big difference. The massive star's core temperature rises up to 600 million degrees Kelvin. When it's this hot, the carbon core begins to fuse and convert carbon to magnesium. Eventually the carbon is used up and further gravitational collapse takes place, causing the temperature to rise again, and new nuclear reactions take place. The star produces heavy elements like nitrogen and silicon until the core is mostly iron."

Brunhilde's screen is flickering, as if she wishes to provide a comment, but her audio output is inhibited for the moment.

"Each burning stage takes place at a higher temperature, because the heavier nuclei contain more protons, which more strongly repel the positively charged alpha particles that approach them. Each stage also takes less time than the previous one. In a 20-solar-mass red supergiant, helium fuses into carbon in a million years. The carbon burning that produces neon and magnesium, takes less than 100,000 years. The last step, the conversion of silicon to iron takes place in a week. Iron is the most tightly bound of all atomic nuclei. In less than a tenth of a second the core, about the size of Earth, compresses further, and the gravitational energy released causes the star to blow up." Bob reactivates Brunhilde. "Brunhilde, display core of massive star." Brunhilde breathes hard for a moment and then displays figure 8.5. "There is a huge explosion as the star becomes 100 billion times as luminous as our Sun. Of course, that's only the visual part of the energy. Most of the energy released in a supernova can't be seen by your eyes. Neutrinos are also released."

Bob walks past the trilobite tank and stands before a grove of wavering kelp that form a sheltered, sun-dappled underwater forest. A few of the plants seem to have small gas bladders that lift the kelp plant toward the light at the top of the tank. The thick canopy of brown and green looks so peaceful that Bob feels like diving in.

"Are all supernovae the same?" asks Miss Muxdröözol. Bob watches her as her earrings cast a tangled, golden reflection in the kelp forest.

"There are two main types of supernovae—those that occur for a single massive star and those that occur because of *mass transfer* in a binary system. A star that might not ordinarily have enough mass to explode may gain mass as it grabs matter like hydrogen and helium plasma from its companion, a process known as accretion. For example, if the recipient is a white dwarf, it may explode when sufficient hydrogen has been deposited on the dwarf's surface to initiate a burst of surface fusion. Novae are common. If enough mass is transferred onto the white dwarf so that it has more than 1.4 solar masses, there may

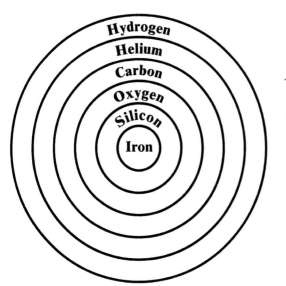

Figure 8.5   The cosmic onion.
Schematic diagram of a massive
star's core minutes before the star
explodes as a supernova.
The iron core is inert, but the other
layers are burning and producing
heavy elements. The outer portion
of the star, not shown here, is
unburned hydrogen.

be a supernova explosion and the white dwarf will collapse through its electron degeneracy into a neutron star. We'll talk about neutron stars soon. In any case, most theoreticians believe that supernovae involving white dwarfs blow the white dwarfs to smithereens except under rare conditions."

Miss Muxdröözol stares at the kelp forest. "Does the heat of explosion create even heavier elements?"

"As I said, stars that are about eight times more massive than our Sun explode when there is no longer enough fuel for the core's fusion process to create an outward pressure combating the inward gravitational pull of the star's great mass. When the iron core is being crushed by gravity the temperature rises to over 100 billion degrees. The repulsive force between the nuclei overcomes the force of gravity, and the core recoils out from the heart of the exploding star. As the shock wave encounters material in the star's outer layers, the material is heated, fusing to form new elements and radioactive isotopes. The shock then propels the matter out into space. Very heavy elements, like gold and lead, are produced in the extremely high temperatures and neutron flux of a supernova explosion. The next time you look at the golden ring about a friend's finger, think of supernova explosions in massive stars."

Miss Muxdröözol plays with her golden earrings. "You mean all the gold I wear is from supernovae."

Bob nods. "Whenever I'm in a museum viewing gold artifacts from ancient Egyptian, Minoan, Assyrian, and Etruscan artisans, I think of supernovas. It's amazing to think that if there were no supernovas, Earth's cultures would have had no single material that was universally accepted in exchange for goods and services. Without the supernova, how would the Spanish discovery of the Americas in the 1490s been affected? Without gold, there wouldn't have been slave labor in the gold mines and the looting of Indian palaces, temples, and graves in Central and South America. There wouldn't have been the huge influx of gold unbalancing the economic structure of Europe." Bob pauses, "Of course, I'm getting carried away focussing on just gold. All heavy elements would never have existed."

Cain, the ant with the multicolored abdomen, approaches Bob. "You must be hungry. Please chew on this." The ant hands Bob, Mr. Plex, and Miss Muxdröözol a foot-long piece of kelp.

Mr. Plex immediately sinks his diamond teeth into the thick frond and makes a sickening chewy sound. "Chewy," he says.

Ishmael folds his multiply jointed leg and begins to eat his portion of kelp, which looks like it has been shredded beforehand. The shreds wound around one another like braids of hair. "We do not eat meat on our world. At least, not anymore."

"Delicious," Miss Muxdröözol says as her two long flat tongues lick her lips, wiping bits of kelp.

Bob looks back into the kelp tank and imagines himself sailing on a blue-sapphire sea with Miss Muxdröözol, alone, with no distractions. Bob sighs. "Anyway, once the gold, carbon, and other nuclei are cast off into space by the supernova, they become available again when new stars and planets are formed. Our own Sun was formed from interstellar gas and dust enriched with these elements. Today there are more light elements in the Universe than heavy ones like gold, because the heavy elements are produced only during the short interval of supernova explosions."

A noise comes from a faraway hallway. It sounds like bowling balls rolling down a long alley. "Please move close to the walls," says Ishmael. The Nephilim nod.

Several seconds later spherical creatures, about 2 feet in radius, roll down the corridor. They have comical, elongated eyes and mouths stretched in an exaggerated smile that never fades. As they move, they sometimes remind Bob of gleaming wheels with eyes in their rims.

Bob watches them as they roll away into the distance. He chooses to ignore the spherical creatures for the moment. "Mr. Plex and Miss Muxdröözol, you may be interested in the many crushed remains of stars. For example, we've talked a little about white dwarfs and neutron stars. We also already talked about how old stars contract due to gravity after their fuels run out. For relatively small stars, the *Pauli exclusion principle* keeps the electrons in a star sufficiently separated to prevent the star from contracting further after it has spent its fuel. In other words, the electrons counteract the crushing gravitational force. However, for stars more than about 1.5 times the mass of the Sun, a mass known as the *Chandrasekar limit*, this repulsive force would not be enough to stop stellar collapse."

One of the rolling creatures returns and listens attentively to Bob's lecture. Bob is beginning to feel important given the diversity and number of his audience: Miss Muxdröözol, Mr. Plex, two Betelgeuse ants, a Roller, Brunhilde, and two Nephilim. He hopes that more visitors at the Nephilim's zoo will come to listen. He raises his voice and shifts into a professorial tone. If only he had his pipe and jacket with leather patched elbows.

"Let me give some examples. Here's a review. If a star is about one solar mass or less, electron repulsion halts complete gravitational collapse, resulting in a stable graveyard state called a white dwarf. A white dwarf is maintained by the exclusion principle repulsion between the electrons in matter. White dwarfs have radii of a few thousand miles, and we can observe a large number of white dwarf stars in the heavens. One of the first to be discovered is a star orbiting around Sirius, the brightest star in the night sky."

The Roller tumbles forward. Its grin seems to be perpetual.

"Aside from white dwarfs, there is another possible graveyard state for stars. Although it's rarer than a white dwarf, stars between 1.4 and about 5 solar masses can explode to create neutron stars when they die. Neutron stars have radii of only around 10 miles, yet they retain all of the mass of the star's core—at least 1.4 times the full mass of the Sun. A neutron star is formed when the core of a supernova collapses inward and becomes compressed together. Neutrons at the surface of the star decay into protons and electrons. Stars with masses greater than about 5 solar masses will become black holes"[6]

Bob starts walking through the Nephilim zoo as his eclectic retinue follow closely behind. Soon another creature comes to listen to Bob's lecture. It has the beautiful warm eyes of Robert Redford, but unfortunately that is where the similarity ends. Above its eyes is a deep fleshy head shield shaped like a helmet. His body consists of seven segments diminishing in size toward his rear, which is about the size of Bob's little finger. Two very long whiplike extensions attach to the last segment and protrude from the creature's ultra tiny rump.

"They're called Hentriacontanes," Ishmael whispers to Bob.[7] "They're among the greatest philosophers in this part of the Universe."

"Wonder what they're smoking?" says Miss Muxdröözol. Several Hentriacontanes wander closer, puffing on fronds of some kind of vermillion weed. Other Hentriacontanes ramble about in apparent boredom. Most circle round and round a frosted stalagmite of greenish ooze. Bob can tell they have been doing this for a long time as their repetitious motions have worn a circular trench into the floor.

"They seem mentally ill," Miss Muxdröözol says to Bob. He nods and begins to wonder. *How can the ants consider them great philosophers?*

Bob continues. "White dwarfs maintain the strongest resistance against gravity, black holes the weakest, neutron stars somewhere in between. An extraordinary material called *neutronium* form neutron stars. Neutronium is so dense that a chunk the size of a thimble would weigh about 100 million tons. This means that a cubic centimeter of neutronium, the size of a sugar cube, has the mass of 100 billion elephants. The Sun, if squashed into a neutron star, would be only a few miles across, the earth just a few inches. The density of neutronium is 10^{14} times greater than ordinary solid matter."

A Hentriacontane comes forward and stares at Miss Muxdröözol.

"Hello," Miss Muxdröözol nervously says as she steps back.

"I have a small gift for you," says the Hentriacontane. The creature displays a vapid insectile grin and offers Miss Muxdröözol a long-stemmed, luminous blossom.

"Maybe you shouldn't accept the gift," Bob says, remembering that the German word *gift* means poison. Miss Muxdröözol takes it anyway.

"I'm honored by your thoughtfulness," she says.

Bob shrugs and continues. "*Pulsars* are also graveyard states of stars, and we believe they are rapidly rotating, highly magnetic neutron stars. They send out steady radio waves that arrive as pulses due to the star's rotation. The pulses are at intervals ranging from milliseconds to several seconds. The fastest millisecond pulsars spin anywhere from 600 to 900 times per second! We've detected hundreds of pulsars. Some actually emit regular bursts of visible light, X-rays, and gamma radiation as well."

"But what causes the radiation pulses?" asks Miss Muxdröözol as she stares into the glowing flower.

"Neutrons at the surface of the star decay into charged particles such as protons and electrons that enter a strong magnetic field surrounding the star. The field rotates along with the star. Accelerated to high speeds, the particles give off electromagnetic radiation released as intense bursts from the pulsar's magnetic poles. Brunhilde, show magnetic poles." Figure 8.6 appears on the flexscreen.

"You can see from this figure that the neutron star has two axes, one about which it rotates and another that corresponds to the magnetic poles. As the star spins, it shoots out radiation like a police car's rotating lights."

"Bob, you mentioned black holes. They seem very mysterious."

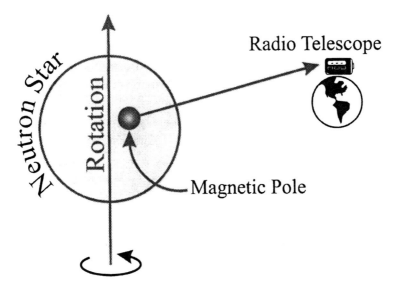

Figure 8.6   A spinning neutron star (pulsar) appears from Earth to flash with a bright pulse of radiation that matches the period of its rotation. Charged particles near the star's surface rotate with the magnetic field and give off electromagnetic radiation.

Mr. Plex grins. "Actually, I did a lot of exploring near black holes when I was younger.[8]

Brunhilde redisplays Figure 8.4. "Black holes can result from supernova explosions that leave behind a massive core," says Brunhilde.

"Yes, another variety of huge black holes exist. We know these cosmological hells truly exist in the centers of some galaxies. These galactic black holes are collapsed objects having millions or even billions of times the mass of our Sun crammed into a space no larger than our Solar System. The gravitational field around such objects is so great that nothing, not even light, can escape from their tenacious grip. Anyone who falls into a black hole will plunge into a tiny central region of infinite density and zero volume, and time will end.[9]

The Hentriacontane who gave Miss Muxdröözol the flower suddenly sprays her with a foul smelling substance that gushes from the pores of its body like oil from a well.

"Hey!" Miss Muxdröözol screams as the oleaginous oobleck drips from her clothes. The Hentriacontane seems startled.

"I'm sorry if I offended you, madame. Isn't it customary for your species to spray pheromones onto new acquaintances of the opposite gender? I am male. Are you not female?"

"Our pheromones are slightly more subtle," she says.

Bob takes a handkerchief out of his pocket and hands it to Miss Muxdröözol. "As some historical background, black holes come in many shapes and sizes. Just a few weeks after Albert Einstein published his general relativity theory in 1915, German astronomer Karl Schwarzchild made exact calculations of what is now called the *Schwarzchild radius*. This radius defines a sphere surrounding a body of a particular mass. Within the sphere, gravity is so strong that light, matter, or any kind of signal cannot escape. In other words, anything that approaches closer than the Schwarzchild radius will become invisible and lost forever. For a mass equal to our Sun's, the radius is a few kilometers. For a mass equal to the Earth's, the Schwarzchild radius defines a region of space the size of a walnut."

Bob stares at his audience. "Sometimes I dream of disappearing from our Universe as I fall close to the speed of light into the tiny central singularity where all the laws and properties of our physical Universe shatter, where gravity turns time and space to subatomic putty, and God divides by zero . . ."

Bob spends the next half hour talking with the Hentriacontanes. They tell him that when they are not studying religion, they spend much of their days investigating a computer simulation called MonkeyGod.[10] The computer program generates artificial universes using different values for four physical constants: the masses of electrons and protons and the strength of electromagnetic and strong nuclear interactions. Most of the simulated universes could never support life, but a few permit the formation of stars.

Bob wishes he could spend a few hours studying the game. Apparently status in the henotheistic Hentriacontane society is based on the prowess with which a Hentriacontane simulates universes. The size of the two whips protruding from an individual's rump is proportional to the individual's mastery of the game.

"Bob," says Miss Muxdröözol, "how many black holes existed in our Galaxy back in the twenty-first century?"

"No one was sure. Dr. Robert Wald once suggested that supernovas, the exploding, dying stars, occurred in our Galaxy at the rate of several per century. If a high fraction of them result in the formation of a black hole, there could have been be as many as one hundred million black holes in our Galaxy. Stephen Hawking believed that, in the long history of the Universe, many stars must have burned all their nuclear fuel and collapsed. The number of black holes, according to Hawking, may have been greater than the number of visible stars, which I told you totaled about one hundred thousand million in our Galaxy alone. Now that we are so far in the future, certainly there are more black holes than visible stars."

Bob is reaching the climax of his lecture on stars. One of the Rollers bounces up and down in anticipation. The Hentriacontanes wave their whip tails as they beckon him to continue. Ishmael taps his leg.

"Bob," says Miss Muxdröözol, "I've been thinking about how heavy elements are synthesized in stars. But couldn't some of these heavy elements have been made in the intense heat and pressure of the Big Bang? Maybe if the Big Bang made carbon, we wouldn't need stars for complex materials."

"No, stars are the crucibles of the heavy elements, and they have several advantages over the Big Bang for producing the heavy elements. For one thing, the core of stars actually compresses matter far more tightly than the Big Bang did when element building might have occured.[11] Nuclear reactions happen more quickly at high density because nuclei bang into one another more often when they are close together. Stars also stay very dense and hot for millions or even billions of years. The explosion of the Big Bang was quite rapid and the Universe cooled so quickly that the conditions for nucleosynthesis—the fancy term astronomers use for building of atoms—existed for only the first ten minutes of the Universe's life. On the other hand, the long life of stars allows various nuclear processes to take place."

Bob finally finishes, and the crowd grows silent. Mr. Plex and Miss Muxdröözol have learned a great deal about stars. Bob hopes they feel a sense of completion and share his sense of wonder about these magical furnaces suspended like little diamonds in the vast deserts of space.

Bob approaches Ishmael. "Now it's time for you to give us some answers. Why have the Nephilim brought us here?"

"We can only speculate," says Ishamel. "We believe the Nephilim wish us to create an intergalactic nursery for creatures of high intelligence. If humans and other species are raised together from birth, they will learn one another's languages and habits. This will allow them to pool their collective knowledge so they can formulate a way to avoid the heat death when all the stars in the Universe eventually burn out and the void is filled with perpetual darkness."

"Heat death?" Mr. Plex says. "You mean when the density and temperature of the Universe keeps dropping?"

Bob puts his hands on his hips. "Why don't you ask God to solve this problem?"

"We asked, but there is no reply. We don't want a heat death to happen. We must learn more. In order to increase our intelligence, we will swap the brain hemispheres of infants of different species. They'll all be hybrids with new ways of looking at the Universe. Our preliminary research indicates that they will think new thoughts and contemplate the fabric of reality in ways we can never fathom. We have the technology to fuse the brains, but not the intelligence to think their thoughts."

"That's sick," Miss Muxdröözol says. "It would be morally indefensible!"

Ishmael shakes his head. "The Nephilim will teach the children how they can best use their new brains. It would be unethical not to give children the expanded ways of seeing the Universe." The ant pauses. "If a child were born with eyes that saw no colors, wouldn't it be cruel to withhold treatment that allowed them to see color?"

Miss Muxdröözol snaps her fingers. "That's not the same."

"We've no right to deny our children species access to the wealth of new ideas," says Ishmael. "Think of this as the next step in your evolution. Think of the new philosophies and ideas we would develop. The children will be happy. They will still love their parents. They will play with friends. They'll just see more of reality. And they may help us save the Universe."

A Hentriacontane nods. "Miss Muxdröözol," he says, "a year ago you could never have contemplated falling in love with a humanoid like Bob. But now, I believe, you can. You have grown. Think of this new experiment with brains as the ultimate merging of God's creatures. The Nephilim are wise. They once helped us build structures around some of the warm dwarf stars to capture their radiation. Despite the Nephilim's bizarreness, I trust them."

Bob looks at the insectile Hentriacontane. There is a bit of irony in a creature like this, surrounded by so much strangeness, calling another creature "bizarre."

A Roller comes close to Bob's foot. "I am sometimes asked *why* the Nephilim are watching us. What is their purpose? Some flavors of quantum mechanics suggest that a conscious observer is necessary to bring subatomic events into concrete reality. I have my doubts that this applies to macroscopic objects, but one theory is that the Nephilim are observing us to make us real. We need a conscious observer to collapse the waveform and validate our existence. The Nephilim have been making humans real since before the time humans were conscious, during the pre-australopithecine era—that is, before the last half of the Late Miocene epoch and the beginning of the Pleistocene epoch. All humans also should have many watchers, just for safety, so that they do not disappear. Your androgynous, android bartender Hieronymus was one of your watchers."

"What?" Bob exclaims, stepping back. "That's insane!"

"The Copenhagen interpretation of quantum mechanics says that the only real things are observed things. God is the supreme observer, who, by watching all particles, converts their quantum potentials into real and actual states."

Bob thinks that the Roller is spouting a lot of nonscientific gibberish. But one thing is certain—he does feel closer to Miss Muxdröözol. The adventure was certainly worth the hardships. Anyone who hears about Bob's journeys

would certainly be enthralled by his descriptions of the exotic places and creatures, by his ability to adjust to adversity, by his humor and incisiveness, but
above all, by the realization that to understand his world, he had to make himself vulnerable to it so that it could change him.

Bob suddenly claps his hands together. "There are just too many questions.
Why don't the Nephilim themselves answer our questions. Where is God in all
of this?"

"Do not question the Nephilim," Ishmael says. "They are so different from
us that their answers are nearly always meaningless, as far as we can tell. But it
is clear they have powers beyond our imagining."

"As for God," the Roller squeaks, "let me see if He is around." The Roller
gives a sharp whistle.

There is a humming sound coming from the next corridor. The lights dim.
The temperature drops. Bob thinks he hears the muttering of a foreign language. It sounds a bit like Hebrew or perhaps Aramaic. A few minutes later, a
creature about a foot tall walks up to Bob. Its head is multicolored, like a rainbow. Its body is trilaterally symmetric, with two legs coming out of each side of
the triangle. The six legs appear to be jointed every few inches, and they are so
reflective they appear to glisten like fire. On each leg are two eyes with lids that
close bottom to top. When one eye closes the other opens. So, at no time are
both eyes closed.

Bob opens one of his mouths wide but then closes it. His other mouth speaks.
"Surely, this is not God."

"God?" the six-legged triangle says. "To establish an unpredictable, creative,
and infinitely evolving Universe, God cannot directly show creatures that He
exists. If He did that, intelligent life would be robbed of their independence.
You, human, must always search for answers on your own." The triangle pauses.
"Why do you think evolution is so painful? Why does the zebra die in blood
and terror as the tiger removes its trachea? Yes, now you know why there is
pain. It is only in a crucible of competition that intelligence can evolve. It was
the only way God could avoid giving away His existence but still produce intelligent beings."

"But who are you?" Bob says. "You are not God."

Out of the mouth of the six-legged triangle spill a number of tiny gems
colored translucent emerald or sapphire. The gems undulate, sprout legs, and
scurry away into the distance.

The triangle steps closer to Bob. "We have played the role of God for a few
years on some worlds. Look at our fiery legs. Recall the words in your own
Bible, Revelation 10:1, 'Then I saw another mighty angel coming down from

heaven. He was robed in a cloud, with a rainbow above his head; his face was like the sun, and his legs were like fiery pillars.'"

"You are the creatures with legs like fiery pillars?"

"Yes, but of course we are not gods. We serve as His messengers, and the Nephilim are our messengers. God has temporarily departed this Universe to create 10^{19} more. In his absence, we are helping to oversee the brain hemispheric merging project. In any case, as I said, He cannot directly intervene, or perhaps He can do so but He cannot let you know He is intervening."

The Nephilim's beards trail along the calcite floor. Their eyes gaze at Bob with an intensity that transgresses the sanctity of the Silurian epoch. Oh, those eyes.

"You have been through much," the Nephilim say. "For the next few days we will create a temporary universe for you, a dream universe. Just make a wish."

This is crazy, Bob thinks. He wants to be away from all of these odd creatures. He needs time to think. He looks into Miss Muxdröözol's eyes.

Instantly Bob and Miss Muxdröözol are walking arm in arm along a simulated, multihued cobblestone road as long black trails of night descend. There is little traffic, except for the occasional shuffle of Betelgeuse ants upon whose backs the bearded Nephilim ride. As Bob walks, a few street lights begin to cast an amber, shadowless glow.

They pass a town center. Bob smiles. The old brick facades of the stores make him feel a wave of nostalgia. Behind the older buildings are newer apartment complexes, then forests and mountains. The immediate downtown area is surrounded by a residential district of haphazardly arranged streets filled with small houses and ancient trees.

The lights from a small store go off suddenly. It is closing time. The parking lot is nearly empty except for a few six-legged triangles parked by an all-night laundry. An occasional Roller passes, and noise comes from a neighborhood bar. In the distance, Bob hears the clattering of chariot wheels on stones.

Bob and Miss Muxdröözol are quiet for the next few minutes. Bob squeezes her hand. "Do you believe in heaven?" he asks.

Miss Muxdröözol suddenly stops and turns to him. "I'm not sure," she says. "Do you?"

"Yes."

"Bob, I didn't think you were religious."

"I'm not. The heaven I believe in is not the typical heaven you hear about in the various religions."

"Then what?"

"Heaven is a particular place," he says gazing at her with one face and then the other.

Miss Muxdröözol shakes her head. "You're not going to suggest that heaven exists on the asteroid with violet grass or in the Nephilim zoo," she says.

"Interesting idea, but no, that's not where I think heaven is."

"Where is it?"

Bob pauses. "Wherever you are, Miss Muxdröözol."

Miss Muxdröözol is quiet for a few seconds. Then she pulls Bob to her and kisses one of his faces. They continue walking in a peaceful silence.

Bob holds her close as a wind stirs the treetops. The breeze feels wonderful. They pass under a large oak tree, and the air is suddenly fragrant with scents of the nearby forests. One of the branches on the oak seems to move ever so slightly in the direction of the two lovers as they passed beneath it. On each one of the trees' leaves is a single unwinking Nephilim eye. The eye gazes at Bob and Miss Muxdröözol until they fade into the distance. Then the branch swivels back to its original position, pointing toward the forest—a wilderness of violet grass, the tips of which glitter like miniature supernovae.

Some Science Behind the Science Fiction

Bright star, would I were steadfast as thou art —
Not in lone splendor hung aloft the night,
And watching, with eternal lids apart,
Like nature's patient sleepless Eremite,
The moving waters at their priestlike task
Of pure ablution round earth's human shores,
Or gazing on the new soft fallen mask
Of snow upon the mountains and the moors . . .

— John Keats (1795–1821),
   "Bright Star, Would I Were Steadfast"

When I was eight years old, I asked my father what would happen when the stars no longer shine. I had heard a rumor that eventually all the stars in the sky would burn out and new ones would not form.

My father had no idea, so I had to wait a few decades until I learned the answer. In about a hundred trillion years, the last generations of stars will have been born, and the few remaining red dwarfs will die. Let's digress for a moment and discuss red dwarfs before talking about the end of the Universe. We've mentioned red dwarfs a few times in this book without giving a detailed description of what they are.

Red dwarf stars have masses between 0.08 and 0.4 solar masses. Like the Sun, they are Main Sequence stars, meaning that they are in the first and longest phase of their existence, burning hydrogen in their cores. When they finish burning their hydrogen, these stars die and become white dwarfs. The temperature inside the red dwarfs never gets hot enough for helium fusion. You can compare red dwarfs to other stars by looking at the Hertzsprung-Russell (H-R) diagram in figure 5.1. Note that red dwarfs are much cooler than white dwarfs, the evolutionary endpoints of stars like the Sun.

One of the most fascinating aspects of red dwarfs is their longevity. The Main Sequence lifetime of a red dwarf is about a hundred billion years. Since this is older than the age of the Universe (which is 10–15 billion years), red dwarfs have not reached the white dwarf stage.

When you look up into the night sky, you never see red dwarfs. They're faint because of their small size and low temperature. Only the nearest red dwarf stars, up to about 90 light-years away, are visible to even the most powerful earth-based telescopes. Proxima Centauri, a member of the Alpha Centauri triple star system, is a red dwarf, and we've discussed that it is the closest star to the Earth. Despite its proximity, it is a hundred times too faint to see with the naked eye. In contrast, we can see many larger stars using our eyes, even when these stars are thousands of light-years away.

Astronomers have found about 450 young red dwarf stars within 80 light-years of the Earth. This means that there are more red dwarfs closer to Earth than all the other types of stars combined. Some astronomers suggest that red dwarfs are the most common type of star in the Universe.[12] Others suggest that even cooler brown dwarfs are the most common star, but they are so dim they are more difficult to detect (figure 8.7).

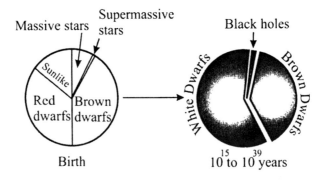

Figure 8.7   Star distributions as pie charts. At left is one proposed distribution of stars born in different ranges of mass. At right is a distribution of stars in the far future (10^{15} to 10^{39} years after the Big Bang). [Star distribution data from: Fred Adams and Greg Laughlin, *The Five Ages of the Universe* (New York: Free Press, 1999), 76.]

Dwarf stars are relatively small stars. By this definition, our Sun is a dwarf star. Dwarfs can be up to 20 times larger than our Sun and up to 20,000 times brighter.

Table 8.1	
Some Dwarfs I Have Known	

Star	Characteristics
Yellow dwarfs	Small, Main Sequence stars. Example: the Sun
White dwarfs	Small, hot, white, dim stars, about the size of Earth. Smaller stars (like our Sun) eventually become faint white dwarfs off the Main Sequence. Nuclear fusion no longer occurs, but the stars possess sufficient heat to be visible using telescopes. These hot, shrinking stars have depleted their nuclear fuels and slowly evolve into cold, dark, black dwarfs. The companion of Sirius is a white dwarf. White dwarfs are not visible to the naked eye but are believed to be common, accounting for 10 percent of all stars in the Galaxy.
Red dwarfs	Small, cool, very faint, Main Sequence star. Surface temperature is less than 4,000 K. Some speculate that these are the most common type of star in the Galaxy. The nearest red dwarf is Proxima Centauri. Red dwarfs are not visible to the naked eye and are about 0.08 to 0.43 solar masses.
Brown dwarfs	Brown dwarfs are star-like objects whose masses are too small to permit nuclear fusion to occur at their cores. (The temperature and pressure at its core are insufficient for fusion.) Brown dwarfs are very abundant in the Universe and are about 0.01 to 0.08 solar masses.
Black dwarfs	Cooled-off white dwarfs. The final stage in the evolution of stars of roughly one solar mass. Black dwarfs are not massive enough to permit nuclear reactions.

Note that if you could stand on the Sun you would weigh only triple what you do on Earth, because even though there is a 300,000-fold increase in mass, the Sun also has a 100-fold increase in radius. However, a white dwarf's radius is so small that a human weighs 8,500 times more than on Earth.

Returning to the question of the eventual demise of all starlight, once the long-lived red dwarfs have died the Universe will be quite a different place than it is today. In 10^{15} to 10^{39} years after the Big Bang (we are currently at around 10^{10} years), the hydrogen-burning stars have given way to an immense grave-yard of stars like brown dwarfs, white dwarfs, neutron stars, and black holes (figure 8.7).[13] Sometimes there is a rare spark of light when two brown dwarfs collide to create a new low-mass star, which will live for trillions of years. There will also be an occasional supernova explosion when two white dwarfs collide. Other than this, it is a mostly dark Universe.

What is the ultimate fate of the Universe, and what life might survive during the final days? Let's first review what the Universe will be like in the far future. Fred Adams and Gregory Laughlin, two astrophysicists, wrote a fascinating article in a 1997 *Review of Modern Physics*[14] describing the birth and death of the cosmos, beginning a million years after the Big Bang and ending an incredible 10^{100} years later. In the current scientifically accepted scenario, our present star-filled cosmos will eventually evolve to a vast sea of subatomic particles as stars, galaxies, and even black holes fade.

The death of the Universe unfolds in several acts. In our current era (10^6 to 10^{14} years after the Big Bang), the energy generated by stars drives astrophysical processes. Even though our Universe is 10 to 20 billion years old, the vast majority of stars have barely begun to twinkle (via fusion). Massive stars burn brightly but die fast. Stars as heavy as the sun live for about 10 billion years. Low-mass stars have not even begun to evolve.

In about 10 trillion years, the emissions of the lowest-mass stars will revive fading galaxies, temporarily boosting their brightness. Alas, even these last sur-viving stars will die after 100 trillion years, and star formation will be halted because galaxies will have run out of gas—the raw material for making new stars. At this point, the stelliferous, or star-filled, era draws to a close.

During the second era (10^{15} to 10^{39} years after the Big Bang), the Universe continues to expand while energy reserves and galaxies shrink, and material clus-ters at galactic centers. Black holes in the center of galaxies absorb some of the material. *Brown dwarfs*, objects that don't have quite enough mass to shine as stars do, linger on. (Even though they do not burn hydrogen in a stable manner, brown dwarfs can be visible, radiating a little energy as they burn their small supply of deuterium to helium and then as they convert their gravitational en-ergy into heat.) Gravity will have already drawn together the burned-out remains of dead stars, and these shrunken objects will have formed super dense objects such as white dwarfs, neutrons stars, and black holes. Eventually even these white dwarfs and neuron stars disintegrate due to the decay of protons.[15]

The third era (10^{40} to 10^{100} years after the Big Bang)—the era of black holes—is one in which gravity has turned entire galaxies into invisible, supermassive black holes. Through a process of energy radiation described by astrophysicist Stephen Hawking, black holes eventually dissipate their tremendous mass. This means a black hole with the mass of a large galaxy will evaporate completely in 10^{98} to 10^{100} years. Protons that might otherwise have existed will have decayed by this point in time.

What is left as the curtain closes on the black hole era at about 10^{100} years? What fills the lonely cosmic void besides photons of large wavelength, neutrinos, electrons, and positrons? Could any creatures survive? In these dark days, our Universe will consist of this diffuse sea of tiny particles. There are no protons, neutrons, or traditional atoms. Fred Adams and Gregory Laughlin estimate the future density of positrons (positively charged particles) is about one particle for every 10^{182} cubic meters. To understand how outrageously low this density is, consider that today our entire observable Universe has a volume of 10^{78} cubic meters. This means that the positron density in the Dark Era of our Universe (10^{101} years after the Big Bang) is about one lonely particle within a volume 10^{104} times larger than our current Universe. These numbers characterize the density for a flat Universe that expands forever but slows down as it expands. Other models of the Universe suggest an accelerating expansion, in which case the cosmic density will be even smaller! In these accelerating models, galaxies that we can currently see—except for those in our local cluster of galaxies—will eventually attain the speed of light and vanish from view. Physicists Lawrence Krauss and Glenn Starkman note that within 2 trillion years, well before the last stars in the Universe die, all stars outside our local cluster will be unobservable and inaccessible. (See their November, 1999 *Scientific American* article in endnote 14.) Krauss and Starkman write, "There will be no new worlds to conquer, literally. We will truly be alone in the Universe." Of course, if we ever develop the ability to use hypothetical spatial shortcuts, like wormholes, this limitation may be banished. In any case, the cosmos will expand forever if the density of matter is too low for gravity to stop the expansion.

I always wonder about the possibility of life in this Dark Era beyond 10^{100} years. Certainly, aliens that depend on water and organic compounds have vanished, but there may be a network of structures, spread out over unimaginable large distances, and these organized structures could store information. According to astrophysicist Gregory Laughlin, these structures, made out of whatever materials are available, will have extraordinarily low energy and will unfold very slowly, but in some sense, the structures may always continue to exist in the Universe. Could these structures be living? What would the lives of these "Diffuse Ones" be like?

* * *

In the Bible, we come across the phrase "for a day will be as a thousand years." During the "Final Days," the consciousnesses of the Diffuse Ones (low-energy-structure creatures) arise from a sea of diffuse electrons or other particles. Their thought and communication processes might appear extremely slow to us in our time frame. But the Universe operates on different scales of time: The time frame of the Earth is much slower than that of humans; the time frame of an insect that lives only 24 hours is much faster than for humans. Similarly, it would not matter to these beings that their thinking, evaluating, and communicating processes were extremely slow by our standards. Their time frame would be proportional to their speed of thought. The Diffuse Ones wouldn't care that it took them a million years to scratch their "noses" or to wait for their toast to pop.

The lives of the Diffuse Ones might be primarily ego-intensive. During the Final Days there would be little outside stimuli to which they could react, and they would look inside themselves for entertainment, ideas, and excitement. They may form complex social groups if communication with other, faraway Diffuse Ones is possible.

Even if they progressively slow down, they could run a virtual world in their minds much like a computer runs a program, and the Diffuse Ones would not perceive any slowdown. This means that although the physical Universe is a black emptiness of electrons, neutrinos, and leptons, a rich virtual Universe could unfold from within.

It is possible that the Diffuse Ones would realize that in the past there were accelerated ways of thinking. To overcome their limitations, they could evolve further in the post-black hole era and choose to organize small pockets of the Universe. They would use their thoughts to gather sufficient particles together and cause a small cosmic egg to form. They could insure their new Universe had an influx of free electrons during its existence. In their new Universe, conditions might be favorable for more rapid thought processes.

Could a Diffuse One have intelligence and consciousness but not be truly "alive?" Some might argue that Diffuse Ones couldn't be alive if they could not reproduce themselves given the limited amount of material available during the Final Days. On the other hand, certainly the nonreproducing bacteria living miles under the Earth are considered life-forms. The definition of life is always difficult. Some people define life as entities that transform energy and replicate. But by using this simple definition, even fire could be considered alive.

These discussions remind me of Frederick Pohl's *The World at the End of Time* in which a number of stars are accelerated at near light speed for millennia until they are far, far away. To the people living around the stars, only a few thousand years passed because of time-dilation effects (time runs slower on high-velocity objects relative to low-velocity objects), but in the meantime the rest of the Universe deteriorates to a very low-energy state with little structure. This is one way of keeping hope (and structure and life) alive in a dying Universe. Perhaps some farsighted beings could toss a few galaxies on a programmed path sending them far away at near light speeds and then bring several back every million years when the rest of the Universe has decayed.

Michael Michaund, former deputy director of the Office of the International Security Policy in the Department of State, and SETI enthusiast, speculates on what humanity (or aliens) could do to avoid eventual destruction: "Organized intelligences of the Universe might avert destructive collapse or dissipation of the Universe by isolating controlled regions of the Universe from the rest of space-time and universal evolution, by transferring themselves to another point in time, or by escaping this Universe, perhaps to another, younger one." (SETI stands for the search for extraterrestrial intelligence. See Walter Sullivan's *We are Not Alone* for more of Michaund's theories.)

Physicist Freeman Dyson in his 1979 *Reviews of Modern Physics* paper examines life in the past and notes that it takes about 10^6 years to evolve a new species, 10^7 years to evolve a genus, 10^8 years to evolve a class, 10^9 years to evolve a phylum, and less than 10^{10} to evolve all the way from the primeval slime to *Homo sapiens*.[16] If life continues in this fashion in the future, it is impossible to set any limit to the variety of physical forms that life may assume. What changes could occur in the next 10^{10} years to rival the changes of the past? Dyson believes that in another 10^{10} years, life could evolve away from flesh and blood and become embodied in an interstellar black cloud, as presented in Fred Hoyle's science fiction story *The Black Cloud*, or in a sentient computer, as presented in Karel Capek's *R.U.R.* In Hoyle's black cloud, a large assemblage of dust grains carries positive and negative charges. The cloud organizes itself and communicates with itself by means of electromagnetic forces. Dyson notes, "We cannot imagine in detail how such a cloud could maintain the state of dynamic equilibrium that we call life. But we also could not have imagined the architecture of a living cell of protoplasm if we had never seen one."

Some physicists suggest that the amount of energy in the Universe will asymptotically approach, but never really reach zero. Life and civilization could get by forever on the tiny residue of energy. However, given a large period of time the energy available to life-forms may become arbitrarily small. At some point life may encounter quantum mechanical effects. What happens when quantization leaves only one energy state? At this point, it may be possible to exist via the vacuum fluctuations where the Universe's overall energy is arbitrarily close to zero but locally there may be variations allowing transient existence—not so different than what we all have now, transient existences.[17]

If you believe that only flesh and blood can support consciousness, then life would be very difficult in the Final Days when the Universe expands and cools. Water and energy would be rare. But to my way of thinking, there's no reason to exclude the possibility of nonorganic sentient beings in the final diffuse Universe. I call these beings Omega creatures. If our thoughts and consciousness do not depend on the actual substances in our brains but rather on the structures, patterns, and relationships between parts, then Omega beings could think. If you could make a copy of your brain with the same structure but using different materials, the copy would think it was you.

And when the stars die, beings like Diffuse Ones could inherit the knowledge and emotions of human civilizations. They may be stepping stones in our final salvation, a bridge to an infinite heaven.

* * *

We've been discussing the possibility of life in a future so distant that stars no longer shine. However, even today, stars are not necessary to support life or produce light. For example, light may be emitted by chemical processes on a planet far away from a sun. A more intriguing idea is the possibility of life on brown dwarfs—the warm planet-like objects far away from suns and therefore without sunlight.

As we've discussed, a brown dwarf is an astronomical object intermediate between a planet and a star, with a surface temperature below 2,200 degrees Celsius. Sometimes described as failed stars, brown dwarfs probably form like stars when interstellar clouds contract into smaller, denser clouds. Unlike stars, however, brown dwarfs do not have sufficient mass to generate the internal heat that in stars ignites hydrogen and creates thermonuclear fusion reactions. Though they generate some heat and some light, brown dwarfs also cool rapidly and shrink. Brown dwarfs look like high-mass planets and may be distinguishable from planets only in their formation mechanism. A brown dwarf is formed directly from a collapsing gas cloud—a stellar process—rather than from the accretion of dust and gas that gives birth to planets.

Incidentally, it's likely that brown dwarfs account for a small portion of the "dark matter" in our Universe. There seems to be ten times more matter in the Universe than astronomers can account for by studying observable stars. For example, galaxies near the Milky Way appear to be rotating faster than would be expected based on the amount of visible matter that appears to be in these galaxies. Other possibilities for dark matter include neutrinos with slight mass, black holes, and exotic subatomic particles that are not detectable by observing electromagnetic radiation.

Returning our attention to brown dwarfs, how could life evolve and survive on bodies with no sunlight? On Earth, there are numerous creatures that live quite happily without light. These include deep-sea and deep-Earth creatures that consume hydrogen sulfide. It's quite possible that the first life-forms on Earth may not have needed light at all. However, although there is no "visible" light available to life on brown dwarfs, warm dwarfs glow brightly in the deep infrared, and this might be exploited by organisms, both for vision and photosynthesis—the formation of carbohydrates by plants. While photosynthesis as we know it would be impossible without sunlight, a different form of energy capture could take place in the absence of sunlight. Moreover, lightning discharges that may have played a role in chemical evolution on Earth would be present on brown dwarfs to provide an abundant energy source.

The nearest life beyond the Solar System may not be on a planet orbiting a star but on a lonely body not married to any sun. Countless bodies of this sort, possibly with water, are probably scattered throughout the Universe. It's difficult for astronomers on Earth to detect brown dwarfs because they are intrinsically faint objects. However, in 1995, a cool brown dwarf named Gliese 229B was discovered by scientists using the 60-inch Palomar telescope specially fitted with a coronagraph, a device usually used to study stars. Tiede 1, a dim object in the Pleiades star cluster, is also generally accepted to be a young brown dwarf. Numerous other possible brown dwarfs have been discovered since then. A dwarf ten times the mass of Jupiter would produce the right amount of heat for liquid water.

What strange biologies might develop in the absence of light in the violet-to-red range on brown dwarfs or other dark worlds? Creatures could "see" using vibration and electrical sensors such as exhibited by the electric eel and mormyrids—elephant fish that have a larger brain per body weight than man. Creatures in the dark might also sense pressure differentials. As an example, consider the fishes that live in caves on Earth. They can orient themselves, like most other fishes, by their lateral line sense, moderated by the organ that runs along their sides and registers the surrounding environment by sensitivity to its differential pressures. Aliens on dark worlds might develop a very keen sense of temperature and use this for both communication and exploring their environment. While humans can sense gross changes in temperature, some animals on Earth possesses thermal sensors far finer than ours. For example, the mosquito can register differences of as little as 0.002 of a degree centigrade at a distance of 1 centimeter.

What would happen if we could visit aliens on a dark world and shine an ordinary flashlight at them? Perhaps the aliens would fear a bright light, if they could perceive it, making light a symbol of great evil or holiness. Do brown-dwarf priests dream of white light in the same way theologians conjure indescribable visions of God? Can creatures dream of things beyond their sensory capacity?[18]

Various researchers have also discussed the possibility of life in white dwarf atmospheres. Because white dwarfs of higher mass contain large amounts of carbon dioxide, life remains a possibility. The atmospheres of older white dwarfs can be quite cool. Fred Adams and Greg Laughlin write, "Even if the evolution of life in a white dwarf atmosphere is 100 million times less efficient than biological evolution on Earth, the star still has enough time and energy to engender a web of life-forms comparable in scope to the biosphere on Earth today."[19]

* * *

Throughout this book we have seen a veritable zoo of strange stars, from the apparently simple Sun, to giant stars, to bizarre, dense stellar end products. For example, Bob and Mr. Plex discussed pulsars, neutron stars, and black holes. To visualize pulsars, think of a ball the size of a small town spinning completely around every second. Better yet, think of a huge lighthouse sending a beacon into outer space. If the beacon of light cuts across the Earth, we receive a pulse of radiation. If not, we may not perceive the pulsar at all. (A lot of the pulsar's radiation is emitted at radio frequencies detectable by radio telescopes.)

If the angle between the magnetic axis and rotation axis is large, we may see a pulse from the other pole as it goes past us. This second pulse is called an *interpulse*. Young pulsars often rotate the fastest and have the highest energies. Some are seen not only in visible wavelengths but in X-rays as well. The fastest millisecond pulsars rotate at incredible rates, close to 900 times per second. Interestingly, these extraordinarily fast pulsars, known as millisecond pulsars or rejuvenated pulsars, are actually the oldest pulsars by far. They were "spun-up" by mass accretion from a companion star.

The apparent surface of a black hole is called the *event horizon*. It's the limiting sphere at which the strength of gravity stops light from escaping. The size of the event horizon depends only on the black hole's mass. Please don't travel closer to a black hole than the event horizon because nothing, not even light, may escape the huge gravitational pull. The formula for the horizon circumference is:

$$C_h = \frac{4\pi GM_h}{c^2}$$

Here, G is Newton's gravitational constant, 1.327×10^{11} kilometers³ per second² per solar mass. The constant c is the speed of light, 2.998×10^5 kilometers per second, and M_h is the mass of the black hole. As an example, a black hole of 1 solar mass, the mass of the Sun, has a circumference of 18.55313 kilometers. For a value of 303 solar masses, the circumference length is around 5,621 kilometers, about the distance from New York to Los Angeles. This means that a mass 303 times the Sun's mass is crammed into a region of space smaller than the United States of America.

Black holes affect time in strange ways, because someone away from the black hole would observe gravity causing time to slow down for someone near the black hole. Far away from the black hole, where space-time is flat, clocks tick at their normal rates. If you were to move close to a black hole and wave your hands at a constant rate, I would notice your hand waving slower and slower as you moved into regions of increasing gravitational curvature. You wouldn't notice this slow down because your heartbeat and thinking processes are slowed down by exactly the same factor as your wristwatch. Similarly, a clock on the ceiling of your home runs slightly faster than a clock on the floor. The clock on the floor experiences more gravity.

Let's pretend that you are my assistant, and I am sending you near a black hole. During one second of your time near the hole, millions of years could have flown past me. This behavior can be described by:

$$t_2 = \frac{t_1}{\sqrt{1.0 - C_h / C}}$$

Here t_2 is the elapsed time when you hover close to the hole as compared to the elapsed time t_1 far from the hole. This means that if you hovered at 1.000001 times the event horizon circumference, one day for you would mean 1,024 days for me!

Similarly, an observer near a black hole *ages* more slowly than one farther from the black hole. Also, the *proper time* of a clock on the surface of a collapsing star is different from the *apparent time* of the collapse, measured by a distant observer. This is because the surface is accelerating with respect to the distant observer. As Jean-Pierre Luminet and others have noted, the contraction of a star below the Schwarzschild radius happens in *finite* proper time, but in an *infinite* apparent time. We will never be able to see the formation of a black hole. We will see the collapse get slower and slower as light from the star becomes redshifted and fainter. (See my book *Black Holes: A Traveler's Guide* for more information.)

What would happen if you came too close to a black hole? Like an ancient ant trapped in amber, I would see you permanently frozen at the event horizon, and your image would gradually fade. In actuality, your body would pierce the event horizon as you plunged toward the singularity at the center of the black hole. At the heart of the black hole is a place where time and space "stop," and all the known laws of physics break down, a place roughly a hundred billion billion times smaller than an atomic nucleus surrounded by void.

Today, there is significant evidence that black holes are not simply interesting theoretical constructs; however, because black holes are not readily observable, astronomers can only collect data that infer the existence of black holes. For example, in chapter 1 we briefly discussed how Cygnus X-1, an intense X-ray source in the constellation Cygnus, is almost certainly a black hole. The X-ray emission is the prime clue. Matter torn off from a neighbor star forms a swirling disk in which gravitational energy is turned into heat and produces X-rays. By observing the orbit of the visible neighbor star around the "invisible" black hole, scientists conclude that the hole must be at least three times as massive as the Sun and packed into a sphere no more than around 15 million kilometers across. Reducing the companion star to the size of a football, Cygnus X-1 is the size of a grain of sand orbiting a few centimeters above the football's surface.

In 1994, a black hole was identified in the core of the Galaxy M87. Scientists used the Hubble Space Telescope to study M87's inner regions and discovered a previously unknown disk of gas whirling at a speed of 750 kilometers per second (1.2 million mph), about 25 times the velocity at which the Earth orbits the Sun. From the rapid motion of this cosmic whirlpool, scientists estimate that the gas is orbiting a central mass possessing around 2.5 billion solar masses. Moreover, the disk is oriented roughly perpendicular to gas jets that shoot from the center of the galaxy—which is what astronomical theory would predict for the behavior of energy around a rotating black hole. Scientists estimate that the hole measures about 5 billion kilometers in radius. This small size and rapidly spinning surrounding disk rule out most other astronomical objects except for a black hole. Today many astronomers believe that black holes inhabit the center of most large galaxies, including the Milky Way and Andromeda.

Among the most exciting evidence for black holes comes from 10 radio telescopes (collectively known as the Very Long Baseline Array, or VLBA, allowing scientists in 1995 to peer into the spiral galaxy NGC 4258. Researchers measured the swirling motion of a gas disk orbiting the galactic core. From this motion, clocked at 900 kilometers per second, astronomers concluded the core has a density greater than 100 million suns per cubic light-year—exceeding the density of any other galactic center ever measured. This is one of the strongest cases yet for a supermassive black hole in a galactic nucleus.

There is a probable relation between quasars and black holes. Quasars are distant and bright objects. Their tremendous brilliance results from a small area at the quasar's center, and astronomers believe that the energy is generated by gas spiraling quickly into a large black hole. As just one example, in 1995, astronomers monitored how fluctuations in light echo from the center of the active galactic nucleus (AGN) in galaxy NGC 5548. By studying the echoes off clouds of gas whirling around the nucleus at a distance of a few light-days, scientists found support for the idea that a black hole's gravity is the AGN's driving force. Observations made using ground-based telescopes, and satellites including the Hubble Space Telescope, suggest that the center of NGC 5548 is powered by a 20-million-solar-mass black hole.

Incidentally, one way to estimate the mass of a black hole is to observe an object that orbits the hole. The relevant formula is derived from Newton's law of gravitation and from the fact that the speed of a circularly orbiting object is the circumference divided by the period. We have a convenient formula for this:

$$M_h = \frac{C_0^3}{2\pi G P_0^2}$$

M_h is the mass of the black hole. The variable C_0 is the circumference of the object's circular orbit around the hole. P_0 is the period of the orbit, that is, the time required to circle the hole once. G is Newton's gravitational constant, 1.327×10^{11} kilometers^3 per second^2 per solar mass. For example, if you completed one orbit around a black hole in 10 minutes, and your orbital circumference was 4,500,000 kilometers (approximately equal to the circumference of our Sun with a diameter of 4.56×10^9 feet), you would estimate the mass of the black hole to be 303 solar masses, or 303 times the mass of the Sun.

The probable black hole in the Andromeda Galaxy's central bulge has 30 million times the mass of the Sun.

Astronomers have discovered over 30 likely supermassive black holes in our Universe. The main requirement for a galaxy to harbor a massive black hole at its center is the presence of a central galactic bulge. Completely flat galaxies seem to lack large holes. Also, the mass of each hole appears to be roughly proportional to the mass of the host galaxy. In particular, central holes weigh about 0.15 percent of the mass of their hosts. The mass of the hole is also re-

lated to the average velocity of stars within the host galaxy, even in areas beyond the hole's direct influence. No one is sure if the black hole came first and
then determined the mass of the galaxy or if the galaxy came first and determined the mass of the hole.[20]

* * *

Many years ago, I first became interested in the remarkable variety of stars
when I started reading science fiction novels. For example, Donald Moffit's
novel *The Jupiter Theft* describes the Cygnans, a race of intelligent, human-
sized beings living on a gas giant planet orbiting a binary star system. When
one of the stars collapses into a black hole, the Cygnans migrate from their
world using 30-mile-long spaceships, the interiors of which contain huge, artificial forests where Cygnans live. Eventually, the Cygnans enter our Solar System and use pieces of Jupiter as a source of energy. Unfortunately, humans
have great difficulty stopping the Cygnans. Cygnans have lost all interest in
anything unrelated to survival due to long isolation in their spaceship. They
therefore have a total disregard for humans.

In Robert L. Forward's book, *Dragon's Egg*, is famous for describing life on high-
gravity stars, in this case a neutron star. Neutron stars are extremely dense, compact stars composed primarily of neutrons. The neutron star has a mass of a star
but the radius of a small asteroid, so its gravitational field is 70,000 million times
that of Earth. In *Dragon's Egg*, the gravity is so high that the atmosphere is only a
few micrometers thick. Mountain ranges are about a centimeter high.

One can imagine life evolving on a neutron star in the same way as life evolved
on Earth. However, the nuclei making up the biological matter do not have
electrons bound to them, as they do on Earth. Instead, a neutron star's biochemistry depends on nuclear reactions mediated by the strong force of the
nuclei, not on electromagnetic forces responsible for terrestrial chemistry.

In *Dragon's Egg*, the dominant form of life are "cheela," intelligent creatures
with the same biological complexity and a similar number of nuclei as human
beings. Their flat, sluggish bodies (50 millimeter diameter, 5 millimeters high)
consist of complex molecules of bare nuclei. They don't have the strength to
extend themselves more than a few millimeters above the crust because of the
crushing gravity. Similarly, they do not breathe or talk because the "atmosphere"
is only a few micrometers thick. They communicate by strumming the crust
with their lower surfaces. The star's 8000-degree-Celsius surface radiates sufficient long-wavelength "light" to enable cheela to see. To cheela, the surface looks
like a bed of glowing coals. It is never dark, so the life-forms never evolved

sleep. There is no moon, so the creatures have no months. Dragon's Egg does not orbit a star, so they have no year.

Someday we may find life on neutron stars, although it would be stranger than we can imagine. If star creatures did exist, they would probably not discover us. They would find it too difficult to travel in space. The collapsed matter making up their bodies would transform into normal atoms when the creatures lifted off into a region of low gravity, and the creatures would literally blow up. Because their biology depends on strong nuclear forces instead of electromagnetic forces—and nuclear reactions happen faster than chemical ones—star creatures would live a million times faster than us. It would be difficult to communicate with such creatures. It would even be difficult to study them with machines, as described in the scenario with the robotic probe. We would have to communicate with them by computer messages. Even if we liked one another, we could never visit them and they could never visit us. The gravity on a neutron star would destroy us, and our gravity would destroy them. We could only enjoy each other's philosophies from afar.[21]

* * *

Recall that the smiling Roller said to Bob, "The Copenhagen interpretation of quantum mechanics says that the only real things are observed things. God is the supreme observer, who by watching all particles converts their quantum potentials into actual states." Bob thought of this as gibberish; however, there are philosophers and physicists today who actually put forth the controversial suggestion that God's existence is established by the "riddle of quantum observership." Timothy Ferris writes in *The Whole Shebang*, "[There is] the riddle of how the early Universe could have evolved in the absence of observers. The riddle may be 'solved' by invoking God as the supreme observer . . ."[22] Similar concepts are discussed in Robert Sawyer's novel, *Calculating God*.

Although some people may argue a particular submicroscopic event in the quantum world can have no practical impact on our ordinary lives, I can give many examples where this argument is false. Consider a helicopter carrying a package of the deadly botulism toxin. The package is dropped in New York City when a device is triggered by the click of a Geiger counter. According to quantum mechanics, the precise instant of the click is purely random. Hence, a quantum event can certainly have a huge impact on our lives! Similarly, small changes in history, brought about by quantum events, can produce amplified effects through time. Imagine what would happen if Cleopatra had an ugly but benign skin growth on her upper lip. The entire cascade of historical events would be different. A mutation of a single skin cell caused by a random photon of sunlight will change the Universe. This entire line of thinking reminds me of a quote from the New Age writer Jane Roberts: "You are so part of the world that your slightest action contributes to its reality. Your breath changes the atmosphere. Your encounters with others alter the fabrics of their lives, and the lives of those who come in contact with them."[23] In her novel, *Memnoch The Devil*, Anne Rice has a similar view when she describes heaven:

> The tribe spread out to intersperse amongst countless families, and families joined to form nations, and the entire congregation was in fact a palpable and visible and interconnected configuration! Everyone impinged upon everyone else. Everyone drew, in his or her separateness, upon the separateness of everyone else![24]

* * *

At one point in our tale, the six-legged triangle justified the existence of pain by saying, "It is only in a crucible of competition that intelligence can evolve. It was the only way God could avoid giving away His existence but still produce intelligent beings." This is reminiscent of John Brooke's quotation:

> Darwin had illuminated the classic problem of theology: the problem of pain. If competition and struggle were preconditions of the very possibility of evolutionary change, then pain and suffering were the price levied for the production of beings who could reflect on their origins.[25]

John Brooke is Professor of Science and Religion and the Director of the Ian Ramsey Centre at Oxford University. Another wonderful Brooke quotation is: "A deity who could make all things make themselves was far wiser than one who simply made all things."[26]

* * *

Bob also said, "If the Sun were shrunk to the size of a sand grain, the closest star would be a few miles away." This means that if two galaxies gradually collided, the various planets would not be disturbed. Our own Milky Way may collide with the Andromeda galaxy in 6 billion years, just as the Sun is enlarging to a red giant.[27] However, if we were somehow alive at that point in time and looked into the night sky, all we might notice is a gradual doubling of the number of stars. Our Sun will be our undoing, not the inevitable collision of Andromeda and the Milky Way. Due to their gravitational attractions, our entire local group of galaxies will one day be merged into a single metagalaxy.[28]

"Thou didst multiply their descendants as the stars of heaven, and thou didst bring them into the land which thou hadst told their fathers to enter and possess."

—Nehemiah 9:23

"All we are is light made solid."

– Anonymous

"Surely it heard me cry out—for at that moment, like two exploding white stars, the hands flashed open and the figure dropped back into the earth, back to the kingdom, older than ours, that calls the dark its home."

—T. E. D. Klein,
Children of the Kingdom

"It is the glory of God to conceal things, but the glory of kings to search things out."

—Proverbs 25:2

CHAPTER 9

Some Final Thoughts

> *When we look at the glory of stars and galaxies in the sky*
> *and the glory of forests and flowers in the living world*
> *around us, it is evident that God loves diversity. Perhaps*
> *the universe is constructed according to the principle of*
> *maximum diversity, a principle that says the laws of*
> *nature . . . are such as to make the universe as interesting*
> *as possible. As a result, life is possible, but not too easy.*
> *Maximum diversity often leads to maximum stress. In*
> *the end we survive, but only by the skin of our teeth.*
>
> — Freeman Dyson, *Science & Spirit*[1]

Our concept and view of the Universe has changed dramatically over the last four centuries. Back in the time of Galileo (1564–1642), there were many who still believed that the Earth was the center of the Universe. As this *geocentric theory* faded away like a dying flower, new theories evolved that suggested the Sun was the center of the Universe, and when this theory died, many considered the idea that the Milky Way galaxy was the center of everything. Today, we realize that our modern telescopes have only begun to reveal the immense numbers and variety of stars, and we know that the Sun is just an ordinary star in our Galaxy that contains roughly 200 billion stars. The *galactocentric view* of the Universe, which placed our Milky Way at the center of the Universe, turned out to be just another centrism that finally died in the 1920s, when astronomer Edwin Hubble proved that the Milky Way was not the only galaxy in the Universe.[2]

In our observable Universe, we know there is roughly one galaxy for every star in the Milky Way. This fact certainly would have disturbed some of the scientists in the last four centuries and destroyed some of their notions about the heavens. Is it possible that the next four centuries will bring equally radical changes in our cosmological theories? Each of our centrisms died hard, and each demise was aided by new tools, new images, and new maps.[3] One possibility is that our Universe is not the only Universe. I believe that new knowledge gleaned in the twenty-first century about other Universes will destroy the centrism that our Universe is all that there is.

Centuries ago, when humans looked up in the sky and drew groupings of stars—the constellations—they were making maps of the Universe. In fact, all societies have invented constellations.[4] The ones with which we are now familiar were developed so long ago that some of their origin is a mystery. Astrohistorians believe they go back to the Mesopotamia of 2000 B.C. The earliest constellation maps were adopted by the ancient Greeks and then by the Romans who gave them Latin names that we use today. In a sense, the constellations are a record and reflection of human civilizations and their thinking (figure 9.1). The most famous constellations are in the zodiac, a set of 12 constellations that lie along the ecliptic, the plane of the Earth's orbit and of the Sun's apparent annual path. Here we see Leo the Lion, Taurus

Figure 9.1   A drawing of a constellation Taurus from Johann Bayer's *Uranometria* (Augsburg: 1603). Bayer used Greek letters to designate the brightness of stars. The bright star in the Bull's eye became alpha Tauri, just like the brightest star in the constellation Centaur became our familiar alpha Centauri.

the Bull, Cancer the Crab, and others. Outside the zodiac we see constellations like the Big Dipper and Ursa Major and Minor, the large and small bears.

Humans have always been interested in making maps of their world, and these maps, like cosmological theories, suggest a progression of centrisms. For example many medieval maps depict Jerusalem at the center of the world. One of my favorite Jerusalem-centric maps is Heinrich Bünting's 1581 "clover-leaf map" (figure 9.2).[5] The three continents of the Old World are shown divided by three seas, Mediterranean, Caspian, and Red, with Jerusalem at the center because of its religious importance. The islands drawn off the coast of Europe are labeled Denmark and England. America, the New World, appears uncertainly in the lower

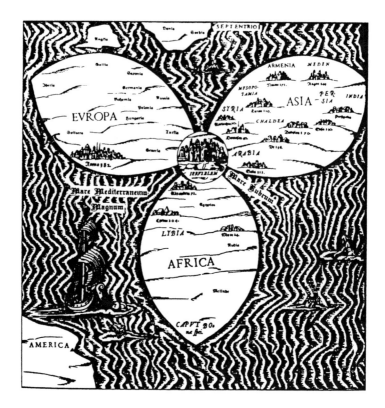

Figure 9.2   A map from 1580 showing Jerusalem as the center of the world.
[Artist: Heinrich Bünting (1545–1606), *Die ganze Welt in ein Kleberblat Welches ist der Stadt
Hannover meines lieben Vaterlandes Wapen,* Woodcut, 26 × 36 cm, in *Itinerarium Sacrae
Scripturae* (Helmstadt: Jacob Lucius Siebenburger, 1581 and later). Donald P. Ryan,
The Complete Idiot's Guide to Biblical Mysteries (New York: Alpha Books, 2000), 149.]

left corner. The explanatory text reads: "as far as the fourth part of the world, namely America, is concerned, which has recently been 'invented,' there is no need here to say more as it is not mentioned in the Holy Scripture." The map appeared in *Itinerarium Sacrae Scripturae,* a travel book based on "the entire Holy Scripture." The book was very popular and appeared in many editions and translations from 1581 to 1774. Bünting was a professor of theology at Hanover in Germany.

It is amazing to consider that the maps we make in the next few decades of the entire Universe will be, in essence, valid for millennia. We live in special times. John Noble Wilford wrote in *The Mapmakers,* "The character and technology of mapmaking may have changed over the centuries, . . . but the potential of maps has not. Maps embody a perspective of that which is known and a perception of that which may be worth knowing."[6]

The images we see coming from modern telescopes encourage us to wonder about our place in this awesome Universe just as maps of the Earth used to lure us from our ordinary lives and towns to the farthest frontiers. Today's cosmic maps stimulate astronomers to explore and ponder the Universe in a manner similar to Marlowe in Joseph Conrad's *Heart of Darkness*:

> Now when I was a little chap I had a passion for maps. I would look for hours at South America, or Africa, or Australia, and lose myself in all the glories of exploration. At that time there were many blank spaces on the earth, and when I saw one that looked particularly inviting on a map (but they all look that way) I would put my finger on it and say, "When I grow up I will go there."

In the sixth century, some people believed that stars were moved by angels who either carried them on their shoulders, rolled them in front of them, or drew them after. Each angel that moved a star carefully observed what the other angels were doing, so that the relative distances between the stars might always remain the same. In the fifth century B.C., Greek philosopher Anaxagoras thought that the Sun was "a red-hot ball of iron not much bigger than Greece."[7] Today we know that the Sun is not made of iron, although scientists thought so until the twentieth century. We also know that the sun is actually about a million miles across—sufficiently large to contain a million Earths. Obviously, we live in an age that is unique for astronomers. We've gone from primitive optical telescopes in the seventeenth century to instruments that now probe the stars beyond the visible spectrum into the radio and X-ray ranges. Our telescopes have left the planet and now float in space. The Hubble Space Telescope allows us to look farther back in time then ever before possible, and in the last few years, we've discovered black holes, brown dwarfs, and planets orbiting other stars. We've come a long way since the 1800s when French philosopher Auguste Comte (1798–1857) wrote, "Never, by any means, shall we be able to study the chemical composition or mineralogical structure of the stars."[8] Despite Comte's prediction, we soon discovered numerous elements—calcium, silver, silicon, carbon, zinc, and more—in the solar spectrum. In the late twentieth century, physicist Richard Feynman emphasized the beauty of such findings when he wrote, "The most remarkable discovery in all of astronomy is that the stars are made of atoms of the same kind as those on Earth."[9]

We have new satellites that measure the background radiation from the Big Bang with increasing accuracy that allows us to test theories using a barrage of new data. We have powerful accelerators probing the deepest mysteries of the atom and amazing ground-based telescopes allowing a wide scan of the heavens. Our computers process information at ever-increasing speeds. It is not an exaggeration to say we live in a special age.[10]

Some have suggested that the biggest and most fundamental discoveries in science and cosmology are behind us and that we have already reached an apex of knowledge. I do not personally subscribe to this belief. In fact, physicist and philosopher Freeman Dyson once likened science to religion in the sense that "science is exciting because it is full of unsolved mysteries, and religion is exciting for the same reason."[11] He believes that to talk about the end of science is just as silly as to talk about the end of religion. Dyson has also often suggested that science and religion are both still close to their origins, with limitless future potential for growth and discovery. One of my favorite quotations from Dyson implies that no matter how much we learn about stars, the Big Bang, and the ultimate fate of the Universe, we will never be bereft of exciting new areas to explore:

> Gödel proved that the world of pure mathematics is inexhaustible; no finite set of axioms and rules of inference can ever encompass the whole of mathematics; given any finite set of axioms, we can find meaningful mathematical questions which the axioms leave unanswered. I hope that an analogous situation exists in the physical world. If my view of the future is correct, it means that the world of physics and astronomy is also inexhaustible; no matter how far we go into the future, there will always be new things happening, new information coming in, new worlds to explore, a constantly expanding domain of life, consciousness, and memory.[12]

Similarly, Isaac Asimov wrote in his autobiography *I. Asimov*, "I believe that scientific knowledge has fractal properties; that no matter how much we learn, whatever is left, however small it may seem, is just as infinitely complex as the whole was to start with. That, I think, is the secret of the Universe."

In chapter 8, we discussed the sensitivity of carbon production to various physical parameters. Carbon is made inside stars and is the basis of life on Earth. The difficulty is to get two helium nuclei in the Sun to stick together until they are struck by a third.[13] It turns out that this is accomplished only because of internal resonances of carbon and oxygen nuclei. If the carbon resonance level were only 4 percent lower, carbon atoms wouldn't form. If the oxygen resonance level were only half a percent higher, almost all the carbon would disappear as it combined with helium to form oxygen.[14] This means that human existence depends on the fine-tuning of these two nuclear resonances. Of these resonances, the famous astronomer Sir Fred Hoyle said that his atheism was shaken by these facts:

> If you wanted to produce carbon and oxygen in roughly equal quantities by stellar nuecleosynthesis, these are just the two levels you have to fix. Your fixing

would have to be just about where these levels are actually found to be . . . A common sense interpretation of the facts suggest that a superintellect has monkeyed with physics, as well as with chemistry and biology, and there are no blind forces worth speaking about in nature. The numbers one calculates from the facts seem to me so overwhelming as to put this conclusion almost beyond question . . . Rather than accept that fantastically small probability of life having arisen through the blind forces of nature, it seemed better to suppose that the origin of life was a deliberate intellectual act.[15]

Physicist Robert Jastrow called this the most powerful evidence for the existence of God ever to come out of science.[16] Other amazing parameters abound.[17] If all of the stars in the Universe were heavier than three solar masses, they would live for only about 500 million years, and life would not have time to evolve beyond primitive bacteria. Stephen Hawking has estimated that if the rate of the Universe's expansion one second after the Big Bang had been smaller by even one part in a hundred thousand million million, the Universe would have recollapsed.[18] The Universe must live for billions of years to permit time for intelligent life to involve. On the other hand, the Universe might have expanded so rapidly that protons and electrons never united to make hydrogen atoms (figure 9.3) If our Universe was only six times smaller than it is today, it would be one thousand times hotter, and no rocky planets would form.[19] If our galaxy contained the same number of stars but happened to be one hundred times smaller, the increased density of stars would lead to high probabilities that other stars would enter our solar system and alter the planetary orbits.[20]

In chapter 7, we discussed how the Sun powers our Solar System by fusing four hydrogen

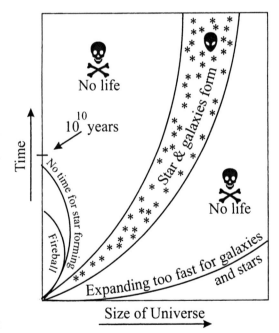

Figure 9.3 "Tuning" of cosmic expansion. Some researchers feel that the expansion of the Universe is tuned to permit the evolution of stars and life. If the universe expands too fast, no galaxies or stars form. If it expands too slowly, the Universe recollapses before stars form. [After Martin Rees, *Just Six Numbers* (New York: Basic Books, 2000), 87, fig. 6.1.]

nuclei (protons) into one helium nucleus (two protons and two neutrons). Because helium is lighter than the four hydrogen nuclei, the mass lost is converted to energy by Einstein's $E=mc^2$. Cosmologist Martin Rees, author of *Just Six Numbers*, points out that the nucleus of a helium atom weighs 99.3 percent as much as the two protons and two neutrons that go to make it (see figure 7.4). This means that the remaining 0.7 percent is released as energy. It follows that the Sun converts 0.007 of the hydrogen mass into energy when it fuses into helium, and the stellar lifetime relies on this number ε (ε=0.007). Subsequent conversion of helium all the way to iron releases only a further 0.001. The strong nuclear force binds the protons in helium and heavier nuclei together so that fusion is a powerful energy source that prevents the Sun from compressing to a graveyard state in 10 million years. If ε were smaller than 0.007, hydrogen would be a less efficient stellar fuel. If the "glue" holding together nuclei were weaker, so that ε=0.006 instead of ε=0.007, a proton will not bond to a neutron, and deuterium (heavy hydrogen, an intermediate molecule in hydrogen fusion) would not be stable. Helium would never be produced if ε=0.006. Without fusion, bright stars would not form in such a Universe. On the other hand, if we set ε=0.008, no hydrogen would have survived from the Big Bang. In our Universe, two protons repel each other so strongly that the nuclear strong force does not bind them together without the aid of one or two neutrons that help glue the nucleus together. However, if ε=0.008, then two protons can bind together directly, and no hydrogen remains in the infant Universe. Without hydrogen, there are no stars or even water. We are lucky that ε=0.007 because we wouldn't be here if ε was 0.008 or 0.006.

Paul Davies has calculated that the odds against the initial conditions being suitable for later star formation as one followed by a thousand billion billion zeroes.[21] Paul Davies, John Barrow, and Frank Tipler estimated that a change in the strength of gravity or of the weak force by only one part in 10^{100} would have prevented advanced life-forms from evolving.[22] There is no physical reason why these constants and quantities should possess the values they do. This led the one-time agnostic physicist Davies to write, "Through my scientific work I have come to believe more and more strongly that the physical Universe is put together with an ingenuity so astonishing that I cannot accept it merely as a brute fact.[23]"

Of course, these conclusions are controversial, and an infinite number of random Universes could exist, ours being just one that permits carbon-based life. Some researchers have even speculated that child Universes are constantly budding off from parent Universes and that the child Universe inherits a set of physical laws similar to the parent, a process reminiscent of evolution on Earth.[24] The Universes that are "successful" from a cosmological-Darwinian perspective are those that produce large numbers of children Universes with long life-times. For example, if we suppose that the central singularities in black holes produce other Universes, as some have suggested, Universes with numerous black holes will be successful. Because many forms of black holes take a long time to form, these Universes will also have time for galactic formation and stellar nucelosynthesis. This means that successful Universes automatically have nearly the right characteristics for the appearance of life forms (figures 9.4 and 9.5). To put it another way, as the cosmological ecosystem evolves, the most common Universes are those that produce large numbers of black holes, stars, and life-forms. If the speculative scenario of evolving Universes describes reality, then our Universe may not be unusual.

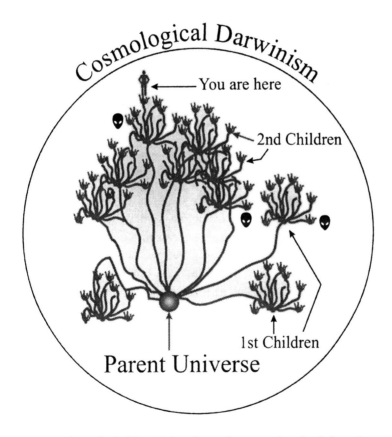

Figure 9.4   Cosmological Darwinism. Researchers postulate that baby universes are spawned from parent universes, and the babies also have babies. According to one theory, the children inherit similar physical laws from their parents, and "successful" universes have a tendency to produce successful offspring. As explained, successful universes are long-lived and have many children and stars, all of which encourages the formation of biological life.

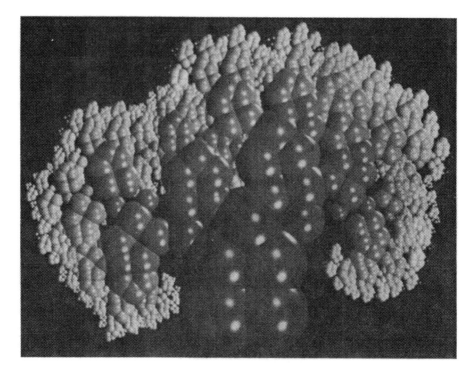

Figure 9.5   Artistic rendition of the cosmic multiverse in figure 9.4.
(Rendition by Clifford Pickover.)

New stars form from clouds of dust and gas when the clouds are able to cool. One of the reasons the contracting clouds can cool is that they contain carbon monoxide and oxygen that radiate away energy. Therefore the presence of carbon and oxygen facilitate star formation, black hole formation, and life formation.[25] We are parasites—intelligent tapeworms feeding off the processes that produce black holes. If black holes create new Universes, we owe our lives not only to the warm suns but also to their blackened corpses.

In this book we have also discussed how stardust, ejected from stars, provides the seeds for planets and life. The dust particles are about the same size as the solid particles in cigarette smoke, and they are spawned gradually when a red giant puffs away its outer layers or more violently when a star explodes as a supernova.[26] Various astrophysicists have speculated wildly on life in the Universe that might be based entirely on unusual physical processes including: plasma life within stars (based on the reciprocal influence of magnetic force patterns and the ordered motion of charged particles), life in solid hydrogen (based on ortho- and para-hydrogen molecules), radiant life (based on ordered patterns of radiation), and life in neutron stars (based on polymer chains storing and transmitting information). It may be difficult to think of these physical processes as being alive and able to organize into complex behaviors, societies, and civilizations. However, when viewed from afar, it is equally hard to imagine that interactions between proteins and nucleic acids could possibly lead to the wondrous panoply and complexity of earthly life—from majestic blue whales and ancient redwoods to curious, creative humans who study the stars. If you were a silicon alien from another star system, and you had a map of human DNA or a list of our amino acids, could you have used it to predict the rise of civilization? Could you have imagined a monarch butterfly, King Solomon's temple, the human eye, the singing of whales, or the smile on a child's face? Would you have foreseen Beethoven, Einstein, Michaelangelo, or Jesus?[27]

In our quest to transcend our lives, which are just miniscule fragments of the vast space-time in our Universe, we have been able to figure out how the stars came into existence. We know why stars shine. We know how they produce the heavy elements necessary for our bones and brains. We also have compelling theories for how the Universe came into existence, by random fluctuations in the quantum foam, governed by the laws of physics. However, we are less sure about where the foam and laws came from. We are less sure about why there is ultimately something rather than nothing. Robert Lawrence remarked in *Closer to Truth*, "Sometimes only silence gets us closer to truth." Timothy Ferris in *The Whole Shebang* similarly believes, "In a creative universe, God would betray no trace of his presence, since to do so would rob the creative force of their independence, to turn them from the active pursuit of answers to mere supplication of God. And so it is: God's language is silence."[28]

John Fowles in *The Magus* sums this up nicely: "There are times when silence is a poem."

> It is while praying that I experience my greatest doubts about God, and it is while looking at the stars that I make the leap of faith.
>
> — Robert J. Sawyer, *Calculating God*
>
> Two things fill my mind with ever-increasing wonder and awe: the starry skies above me and the moral law within me.
>
> — Immanuel Kant

Notes

Twinkle, twinkle, little star,
How I wonder what you are.

— Jane Taylor (1783-1824)

Jane Taylor might have been interested to know *why* stars twinkle. When the starlight passes through the various parts of the Earth's atmosphere—which include the ionosphere, ozonosphere, thermosphere, mesosphere, stratosphere, and troposphere—the light is altered by the different light-bending properties within and between the atmospheric layers. This bending, or refraction, randomly shifts the star's apparent position by a tiny amount, causing us to see a twinkling.

Introduction

1. Rudy Rucker, *The Fourth Dimension* (Boston: Houghton-Mifflin, 1984).
2. *Woodstock* is also sung by Crosby, Stills, Nash, and Young on their album *Deja Vu*. For a wonderful overview of nucleosynthesis in stars, see Marcus Chown, *The Magic Furnace: The Search for the Origins of Atoms* (New York: Oxford University Press, 2001).
3. University of Maine at Machias, "van Gogh," http://www.umm.maine.edu/faculty/gnecastr/vangogh6.htm
    Also see, Bob Brooks, "The Letters from Vincent to Theo," The Vincent van Gogh Information Gallery, http://www.vangoghgallery.com/letters/506.htm
4. Clifford A. Pickover, *Strange Brains and Genius* (New York: Quill, 1999).
5. van Gogh called his seizures "the storm within," and he realized that they were the source of his hallucinations, unprovoked feelings of anger, fear, confusion, and uncontrolled floods of early memories. In one typical seizure, van Gogh saw every room in the house where he had been born, "every path, every plant in the garden, the views in the fields roundabout . . . down to a magpie's nest in a tall acacia in the graveyard." In 1888 while in a hospital, van Gogh painted more expressively than ever before.
6. While working with Dr. Ed Krupp of the Griffith Park Observatory, Albert Boime recreated the night sky of June 19, 1889 at 4:00 A.M. and noticed the striking resemblance to the night sky in *Starry Night*, with the exception that the Moon on that night seems not to have been a crescent but rather a gibbous moon. See Nancy Hathaway, *The Friendly Guide to the Universe* (New York: Penguin Books, 1984), 428. See also, Eames Demetrios, "Vincent van Gogh," http://www.powersof10.com/powers/people/people/people_38.html (Eames Demetrios is creator of the CD-ROM "Powers of Ten™ Interactive." Powersof10.com is an on-line guide, bibliography, and resource for this media.) Another good source is Albert

Boime, *Starry Night: A History of Matter, A Matter of History* (New York: Voyager, 1995). (This is an Interactive media for Mac/Windows CD-ROM detailing the creation of Vincent Van Gogh's most famous work.)

7. Henry M. Morris, "The Stars of Heaven," http://dns1.epcc.edu/faculty/jesseh/imp-010.htm or http://www.icr.org/pubs/imp/imp-010.htm

8. The Pleiades are a cluster of several hundred stars, about 415 light-years from our solar system in the direction of the constellation Taurus. Note that the Japanese name for the Pleiades is Subaru (like the car), and an illustration of Subaru adorns the nameplate and emblem of the cars. In ancient times, the system was known to contain six or seven member stars visible to the naked eye. The Pleiades are mentioned in Homer's novel *The Odyssey* and in the Bible. In Greek mythology the Seven Sisters (Alcyone, Maia, Electra, Merope, Taygete, Celaeno, and Sterope), daughters of Pleione and Atlas, were transformed into stars.

9. Orion is a constellation represented on pictorial charts as the figure of Orion, the hunter in Greek mythology, standing with uplifted club (see figure 5.5). Three bright stars represent his belt and three fainter stars aligned south of the belt represent his sword. The star Betelgeuse, discussed later in this book, is located in Orion. Similarly, the Orion nebula, discussed many times in this book, is located in one of the "stars" in the sword.

10. The Big Dipper, an easy to see constellation in the northern celestial hemisphere, was known to the ancient Greeks as the Bear and the Wagon and to the Romans as Ursa Major (the Great Bear).

11. Martin Gardner, "The Star of Bethlehem," *Skeptical Inquirer*, 23(6): 13–15 (1999). Gardner notes that although Jupiter and Saturn were in close conduction in 7 B.C., they were not so close to appear as one star. Kepler later decided that the Star of Bethlehem was created by God between Jupiter and Saturn when they were close together. Other planetary conjunctions that also occurred around the time of Jesus also have been considered as candidates for the Star of Bethlehem. Gardner comments, "I find it hard to comprehend why conservative and fundamentalist Christians, who believe the Bible's miracles to be actual events, would even try to find natural explanations for what the Bible clearly describes as divine supernatural phenomenon." Through history, the miraculous appearance of a star has seemed to herald the appearance of numerous famous individuals, such as Pythagoras, the mystic geometer, and Krishna, the Hindu savior.

12. "The on-line Islamic Digest," http://islamicdigest.org/articles52.htm

13. Henry M. Morris, "The Stars of Heaven," http://dns1.epcc.edu/faculty/jesseh/imp-010.htm or http://www.icr.org/pubs/imp/imp-010.htm

14. Knowledge Adventure, Inc., "Our home: the Milky Way Galaxy," http://www.letsfindout.com/subjects/space/galaxy.html (Web page now defunct.)

15. There are also brown dwarfs to consider. Brown dwarfs are astronomical objects somewhat between a planet and a star and have a mass less than 0.08 times the mass of our sun and a surface temperature below 2,500 K. (As comparison, the cool red dwarfs are about 3,000–3,400 K). A large number of brown dwarfs would not change how bright the Galaxy appears in optical observations but would change its total mass quite substantially.

    See, William Keel, "How many stars are there in the universe?" http://www.landfield.com/faqs/astronomy/faq/part8/section-4.html

16. Many web sites give interstellar distances. See, for example, John Baez, "Distances," http://math.ucr.edu/home/baez/distances.html. Note that approximate values are used at the larger size scales. A typical equation for computing the maximum radius of the universe looks like:

$$R_{max} = \frac{c}{\sqrt{4\pi\rho_u f}}$$

R_{max} is the maximum radius of the universe, c (3×10^{10} cm/s) is the speed of light, ρ_u (10^{-29} g/cm^3) is the density of the universe, and f (6.67×10^{-8} cm^3/g s^2) is the constant of universal gravitation. See, for example, P. Kittl and G. Díaz, "A simple model of the creation and development of a periodical universe," http://www.cec.uchile.cl/~cabierta/revista/8/simple.html. For related topics, derivations, and assumptions, see P. Kittl, "Some observations on quantum mechanics history, on Planck's elemental cell, on a universe beginning and ending, on mini-black holes, and on a massive binary atom," *Anales de la Sociedad Científica Argentina* (paper in English), 228(1): 8999, 1998. (Address for reprints: P. Kittl, Dept. Mecanica, Universidad de Chile, Casilla 2777, Correo 21, Santiago, Chile.)

17. Einstein's special relativity requires that all objects embedded in space obey the light-speed limit, but it places no restriction on the speed of expansion of space itself. Our universe appears to be expanding, and in fact large portions of faraway space are expanding faster than light. However, there seems to be no way to exploit this spacetime stretching for sending superluminal signals from one part of space to another.

18. We first encounter Mr. Plex and his friend Mr. Veil in my books *Black Holes: A Traveler's Guide* (New York: Wiley, 1996), *Time: A Traveler's Guide* (New York: Oxford University Press, 1998), and *The Loom of God* (New York: Plenum, 1997).

19. In fact, the Earth will be uninhabitable within one billion years. Increased warmth from the Sun will cause carbon dioxide to come out of carbonates at a higher rate than today, which will warm the Earth further and cause more evaporation. The Earth will continue to warm in a runaway "greenhouse."

Chapter 1

1. The Earth is 149,600,000 km from the Sun.
2. Alpha Centauri is actually a triple star, the faintest component of which, Proxima Centauri, is 4.3 light years from the Sun, making it the closest star to the Sun. Alpha Centauri is the star with the largest known parallax. Humans have detected 58 stars within a distance of 5 parsecs from the Sun. These stars include Sirius, Procyon, and Altair.
3. You don't need to know trigonometry to use this formula. However if you are familiar with trigonometry, note that all of the angles used are very small, and I use approximations valid for small angles. Two useful pages on parallax and Friedrich Wilhelm Bessel are:
http://www-groups.dcs.st-and.ac.uk/~history/Mathematicians/Bessel.html
http://www.cyburban.com/~mrf/a_distance.htm
4. Satellite measurements of parallax are useful because satellites are above the Earth's atmosphere, the turbulence of which can limit the parallaxes observable from Earth.
5. William Bixby, *The Universe of Galileo and Newton* (New York: Harper & Row, 1964),71.
6. Ibid., 66.
7. Joshua 10: 12–14 (King James Version).
8. Bixby, 68.
9. Robert Lawrence Kuhn, *Closer to Truth* (New York: McGraw-Hill, 2000), 353.
10. Recent observations of stars orbiting the Galactic center (at distances greater than our Sun's distance from the center) suggest that the Milky Way is close to 150,000 light-years across.

Chapter 2

1. Sara Williams, "The Old Astronomer to His Pupil," in *Best Loved Poems of the American People*, ed. Hazel Felleman (Garden City, New York: Garden City Publishing Co., 1936), 613–614.

2. Electromagnetic waves have electric and magnetic components and are produced by the oscillation or acceleration of an electric charge. Electromagnetic radiation can be represented as a spectrum ranging from high frequency (short wavelengths) to low frequency (long wavelengths). What you and I see with our eyes corresponds to a minuscule section of the electromagnetic spectrum. The electromagnetic spectrum ranges from gamma rays, X-rays, and ultraviolet radiation, to visible light, infrared radiation, microwaves, and radio waves.

3. The Kelvin temperature scale has as its zero point absolute zero, the theoretical temperature at which the molecules of a substance have the lowest energy. This temperature, absolute zero, corresponds to -273.15 on the Celsius scale and to -459.67 on the Fahrenheit scale.

4. See endnote 2 for background on electromagnetic radiation.

5. For a nice web page on spectroscopy, see James A. Plambeck, "Introductory university chemistry: line spectra," http://www.chem.ualberta.ca/courses/plambeck/p101/p01214.htm

6. The following web page has information on many interesting physical constants: "CODATA internationally recommended values of the fundamental physical constants" http://physics.nist.gov/cuu/Constants/index.html

7. Ibid.

8. Frank Tipler, *Physics* (New York: Worth Publishers, 1976), 961–964.

9. For a useful web page on the Rydberg constant and these formulas, see Eric W. Weisstein, "Rydberg constant," http://www.treasure-troves.com/physics/RydbergConstant.html

10. Guillermo A. Lemarchand, "Detectability of Extraterrestrial Technological Activities," http://www.coseti.org/lemarch1.htm. This paper was originally presented at the Second United Nations/European Space Agency Workshop on Basic Space Science, co-organized by The Planetary Society in cooperation with the Governments of Costa Rica and Colombia, November 2–13, 1992, San Jose, Costa Rica — Bogota, Colombia. The paper was republished in *SETIQuest*, 1(1): 3–13. Dr. Lemarchand is coordinator of the META II SETI project in Argentina. In this search for extraterrestrial intelligence, his team is observing the southern hemisphere with a 30-m radiotelescope connected to an 8.4 million channel analyzer with 0.05 Hz spectral resolution.

11. Ibid. See also A.T. Lawton, "Infrared interstellar communication," *Spaceflight*, 13(3): 83–85 (1971); C. H. Townes, "At what wavelengths should we search for signals from extraterrestrial intelligence?" *Proceedings of the National Academy of Science U.S.A.* 80: 1147–1151 (1983); J. D. Rather, "Lasers Revisited: their superior utility for interstellar beacons, communications, and travel," *Journal of the British Interplanetary Society*, 44: 385–392 (1991).

12. Guillermo A. Lemarchand, "Detectability of extraterrestrial technological activities," http://www.coseti.org/lemarch1.htm. See also, Freeman Dyson, "Search for artificial stellar sources of infrared radiation," *Science* 131: 1667–1668 (1959); Freeman Dyson, "The search for extraterrestrial technology," in *Perspectives in Modern Physics* (Essays in Honor of Hans Bethe), ed. R. E. Marshak (New York: John Wiley & Sons, 1966).

13. Freeman Dyson, "Search for artificial stellar sources of infrared radion."

14. Guillermo A. Lemarchand, "Detectability of extraterrestrial technological activities." See also, J. Jugaku and S. Nishimura, "A Search for Dyson Spheres Around Late-Type Stars in the IRAS Catalog," in *Bioastronomy: The Search for Extraterrestrial Life*, eds. J. Heidemann and M. J. Klein, "Lectures Notes in Physics 390" (New York: Springer-Verlag, 1991); V. I. Slysh, "Search in the Infrared to Microwave for Astro-engineering Activity," in *The Search for Extraterrestrial Life: Recent Developments*, ed. M. D. Papagiannis (Boston: Reidel Publishing Company, 1985); N. S. Kardashev and V. I. Zhuravlev, "SETI in Russia," paper presented at the IAA/COSPAR/IAF/NASA/AIAA symposium on SETI: A New Endeavor for Humankind (Washington, D.C.: The World Space Congress, August 30, 1992).

15. Carl Sagan and Iosif Samuilovich, *Intelligent Life in the Universe* (San Francisco, California: Holden-Day, Inc., 1966).
16. Oddly enough, astronomers have found technetium lines around odd stars called S class stars. (You'll learn about spectral classes in the next chapter). S-class stars are among the coolest stars, reddish-brown in color. These stars show zirconium oxide and unusual metal lines such as barium. Further research is required to know for sure if the presence of an uncommon atom in a stellar spectrum is a sign of extraterrestrial intelligence. Another possibility is that advanced civilization may have been continually discarding their dangerous radioactive wastes by shooting them into their suns, and, under certain circumstances, this may be detectable by spectral analysis.
17. Hubert Reeves, *Atoms of Silence: An Exploration of Cosmic Evolution* (Cambridge: MIT Press, 1984); M. Beech, "Blue stragglers as indicators of extraterrestrial civilizations?" *Earth, Moon, and Planets* 49: 177–186 (1990); Guillermo A. Lemarchand, "Detectability of extraterrestrial technological activities."
18. Ibid. (All references in note 17.)
19. Rex Saffer and Dave Zurek, "Blue stragglers in globular cluster 47 Tucanae," http://oposite.stsci.edu/pubinfo/PR/97/35/a.html
20. Reeves, *Atoms of Silence*, 123–124.
21. Clifford Pickover, *The Science of Aliens* (New York: Basic Books, 1999).
22. Clifford Pickover, *The Loom of God* (New York: Plenum, 1998).

Chapter 3

1. Mark Reed, "Discover dialogue," *Discover* 21(9): 26 (September 2000).
2. Table adapted from Dina Moché, *Astronomy* (New York: John Wiley & Sons, 1998), 69.
3. The Corona Borealis can be found between the constellation of Hercules and Boötes. It is visible in the Northern sky from February through September.
4. Bob Halliday and Robert Resnick, *Physics* (New York: John Wiley & Sons, Inc., 1966), 1009.
5. Barry Evans, *Everyday Wonders* (New York: Contemporary Publishing, 1993).
6. Ibid.
7. Bob Herman, "We're going thataway" *Discover* 21(9): 50 (September 2000). Also see, "Calculating the Sun's orbit around the galaxy: a long, strange journey," http://astro.geoman.net/us/actu/astronomie/23/html/solorbite.html. Superclusters (clusters of clusters of galaxies) are the largest gravitationally bound systems observed by astronomers. Our Local Group is a part of the Local Supercluster. Superclusters are located in thin sheets that border empty voids.
8. Richard K. Clingempeel, "Stellar classifications," http://oerlicon.freeyellow.com/./SuperNovae/Classifications.html
9. Ibid.
10. John Gribbin with Mary Gribbin, *Stardust* (New Haven, Connecticut: Yale University Press, 2000), 212.
11. American Astronomy Association, "The History of Women in Astronomy," http://cannon.sfsu.edu/~gmarcy/cswa/history/history.html. See also, Lake Afton Public Observatory, "Women in Astronomy," http://www.twsu.edu/~obswww/wia.html
12. Deborah Crocke and Sethanne Howard, The University of Alabama Department of Physics and Astronomy, "4000 Years of Women in Science: Cecilia Payne-Gaposchkin," http://www.astr.ua.edu/4000WS/GAPOSCHKIN.html
13. Katherine Haramundanis, ed., *Cecilia Payne–Gaposchkin: An Autobiography and Other Recollections,* 2nd ed. (New York: Cambridge University Press, 1996). Also see, John Beaver, "Cecilia Payne-Gaposchkin: In Her Own Words," http://web.physics.twsu.edu/lapo/cpg.htm

14. Jocelyn Bell Burnell, "Cecilia Payne-Gaposchkin," http://www.star.ucl.ac.uk/~hwm/
burnell.htm. See also, Jocelyn Bell Burnell, "Cecilia Payne-Gaposchkin," *The Journal of the
British Astronomical Association* 106(6), 22 (December 1996). (Jocelyn Bell Burnell is professor of physics at The Open University, and one of the first female professors of physics
in the United Kingdom.)

Chapter 4

1. John A. Ruben, Cristiano Dal Sasso, Nicholas R. Geist, Willem J. Hillenius, Terry D. Jones,
and Marco Signore, "Pulmonary function and metabolic physiology of theropod dinosaurs,"
http://cas.bellarmine.edu/tietjen/Ecology/pulmonary_function_and_metabolic.htm. (Also
appears in the January 22, 1999 issue of *Science*, pages 514–516.)
2. Power is always expressed in energy per unit of time. One horsepower is equal to the
amount of power required to lift 33,000 pounds a distance of 1 foot in 1 minute. One watt
equals the power needed to do 1 joule of work per second. There are 746 watts in 1 horsepower.
3. Gianfranco Vidali, "Rough Values of Power of Various Processes (watts)" http://
www.phy.syr.edu/courses/modules/ENERGY/ENERGY_POLICY/tables.html
4. G. Bothun, University of Oregon Physics Department, Electronic Universe Project, "Astronomy hypertext book," http://zebu.uoregon.edu/text.html
5. James Kaler, *Stars* (New York: Scientific American Library, 1998), 140.
6. Hubert Reeves, *Atoms of Silence* (Cambridge: MIT Press, 1984), 44–45, 230–231; Fred Adams
and Greg Laughlin, *The Five Ages of the Universe* (New York: Free Press, 1999), 21–22.
7. As an aside, consider an infinite universe. They sky would be very bright, but not *infinitely*
bright. Beyond a certain distance, the surface areas of the stars appear to touch and make
a shield. Using these assumptions, we obtain a total flux 100,000 times as intense as the
Sun's. See Reeves, *Atoms of Silence*, 230–231, for more information.
   Also note that a hidden assumption in these discussions is that stars have infinite lives,
that is, an infinite fuel supply. Because we know this is not true, Olbers' paradox was never
actually a paradox. For further reading on this subject, see Edward Harrison, *Darkness at
Night: A Riddle of the Universe* (Cambridge: Harvard University Press, 1989).
8. Even if the Universe were static, the fact that the Universe was *born* limits the light of the
night sky to levels acceptable for life. Also consider that stars themselves do not live forever. The average lifetime of stars is too short for light to reach the Earth from very distant
stars. In an expanding universe, the universe is too young for light to have reached the
Earth from distant regions.

Chapter 5

1. One of my favorite authors, Stephen Baxter, also has his characters assess the age of the
universe by monitoring the universe's background radiation temperature. See Stephen
Baxter, *Manifold Time* (New York: Del Rey, 2000), 183.
2. Dina Moché, *Astronomy*, 4th ed. (New York: John Wiley & Sons, 1998), 80.
3. Students for the Exploration and Development of Space, "First direct image of the surface
of a star," http://seds.lpl.arizona.edu/nodes/NODEv5n4-3.html. Here's a nice quote about
Betelgeuse: "Astronomers are drawn to objects that are big, nearby, and bright. That is why

Betelgeuse, a red supergiant star about 430 light-years away, has become a darling. It's our closest red supergiant. The star's enormous body of boiling gas—fueled by nuclear fusion in its core—would extend to Jupiter's orbit if it were substituted for the Sun." (Tracy Staedter, "A favorite supergiant gets even more popular," *Astronomy* 27(9): 30, (September 1999). Posted at Britannica.com, http://www.britannica.com/magazine?ebsco_id=200182)

4. Martha Haynes and Stirling Churchman, Cornell University Department of Astronomy, "Red giants and Supergiants," http://astrosun.tn.cornell.edu/courses/astro201/red_giant.htm
5. Greg Donohue and Jim Ehrmin, the Everett Astronomical Society, "Star light, star bright," http://galaxyguy.bizland.com/KSER200003.htm
6. David Spergel, Gary Hinshaw, and Charles Bennett, NASA, "The Cosmic microwave background radiation," http://map.gsfc.nasa.gov/html/cbr.html
7. James Kaler, *Stars* (New York: Scientific American Library, 1998), 141.
8. Ibid.

Chapter 6

1. John Wood, "Forging the planets: chaos and collision, fire and ice." *Sky & Telescope* 97(1), (January 1999). Reprinted at Britannica.com, http://www.britannica.com/bcom/magazine/article/0,5744,76385,00.html. Long before some new stars begin their fusion reactions, they unleash strong X-ray flares that reach temperatures of 100 million degrees Celsius. For example, newborn stars (proplyds) in the Orion Nebula emit surprisingly hot bursts of X-rays. For more information, see Robert Irion, "X-rays hit the spot for astrophysics," *Science* 290(5498): 1884 (December 8, 2000).
2. Solar neutrinos, which interact very weakly with matter, should also be produced by the nuclear fusion reactions in the Sun. However, scientist's detect much fewer neutrinos than expected, which may suggest that our knowledge of the solar processes that cause the Sun to shine or of neutrinos themselves is incomplete.
3. Dainis Dravins, Luund Observatory, "Solar granules: the solar photosphere in white light," http://nastol.astro.lu.se/~dainis/HTML/SOLAR.html
4. Solar granule information comes from: Amara Graps, Stanford Solar Center, "Solar granulation quiz!" http://solar-center.stanford.edu/cgi-bin/quiz2.pl/granule_quiz.html
5. For more interesting information, see the quiz at: Amara Graps, Stanford Solar Center, "Sunspot quiz!" http://solar-center.stanford.edu/cgi-bin/quiz2.pl/sunspot_quiz.html
6. Space Science News, "A twisted tale of sunspots," http://www.spacescience.com/headlines/y2000/ast29feb_1.htm. See also "Myth: The sun is a ball of hot gas," *Discover* 22(2): 17 (February 2001). Almost all of the Sun is not a gas but rather a plasma, a mixture of charged atoms and the electrons stripped from those atoms. Unlike a traditional gas, plasma holds a magnetic field, which is why solar flares may exhibit looping structures. Magnetic fields through the plasma drive the sunspot cycle and help transport energy from the Sun's interior. Scientists have discovered mammoth rivers of hot, electrically-charged gas (i.e., plasma) flowing beneath the Sun's surface. They have also found features similar to trade winds that transport gas beneath the Sun's fiery surface.
7. Ibid.
8. David H. Hathaway, NASA, "The sunspot cycle," http://science.msfc.nasa.gov/ssl/pad/solar/sunspots.htm
9. David H. Hathaway, NASA, "Sunspot cycle predictions ," http://science.msfc.nasa.gov/ssl/pad/solar/predict.htm
10. Ibid.

11. James Kaler, *Stars* (New York: Scientific American Library, 1998), 126.

12. Space Science News and NASA, "The day the solar wind disappeared," http://spacescience.com/newhome/headlines/ast13dec99_1.htm

13. Ibid.

14. Bill Arnett, "Heliopause," http://seds.lpl.arizona.edu/nineplanets/nineplanets/medium.html#heliopause

15. Steve Mercer, "The Heliosphere," http://www.spacescience.org/ExploringSpace/Heliosphere/1.html

16. Dina Moché, *Astronomy* (New York: John Wiley & Sons, 1998), 105.

17. NASA, "What is a Solar Flare?" http://hesperia.gsfc.nasa.gov/sftheory/flare.htm

18. Gordon Holman, NASA, "High Energy Solar Spectroscopic Imager (HESSI)," http://hesperia.gsfc.nasa.gov/hessi/index.html

19. NASA, "Why study solar flares?" http://hesperia.gsfc.nasa.gov/sftheory/studyflare.htm

20. Marcus Chown, *The Magic Furnace* (New York: Oxford University Press, 2001).

21. Ibid.

22. Ibid., 210.

23. Ibid., 211.

24. Ibid., 213.

25. Ibid., 214.

26. Thomas Hayden, "Curtain call," *Astronomy*, January 2000, 45–49.

27. Kalher, *Stars*, 143–144.

Chapter 7

1. Stephen Baxter, *Manifold Time* (New York: Del Rey, 2000), 192.

2. Robert Sawyer, *Calculating God* (New York: Tor Books, 2000), 63.

3. Ibid. Also see John B.O. Mitchell, "Probabilistic problems in the origin of a habitable universe and the origin of life," http://www.biochem.ucl.ac.uk/~mitchell/origins.html; Paul Davies, "Is there a meaning behind existence?" http://salam.muslimsonline.com/ ~muslimrr/davies.html (excerpts from Davies's 1984 book *Superforce*); George Greenstein, *The Symbiotic Universe* (New York: William Morrow, 1988); Paul Davies, *God and the New Physics* (New York: Simon & Schuster, 1984). There appear to be numerous other physical and mathematical "coincidences" that permit stars to form and life to evolve. For example, if the strong nuclear force were only a few percent weaker, deuterium would not exist, and there would be no nuclear fusion. If the strong force were a few percent stronger, helium would be the dominant element of the Universe, not hydrogen, and there would be few stars. If the charges of a proton and an electron differed by as little as one part in 100 billion, relatively small objects like rocks and people would fly apart. If the charges of a proton and an electron differed by as little as one part in a billion billion, larger objects like the Earth and the Sun would fly apart. There is no logical reason why space should have three large dimensions. Yet if space had less than three dimensions, it is doubtful that complex life forms could have evolved (e.g., complete digestive tracts would cut a creature in two pieces). If space had more than three large dimensions, orbits would be unstable, and there could be no planet circling a life-giving star. (I discuss the effect of spatial dimension on life-forms in my book *Surfing through Hyperspace* (Oxford University Press, 1999). The Big Bang itself is quite odd to say the least. In the absence of other forces, systems tend to collapse gravitationally into black holes. Astronomer Roger Penrose estimates that the odds against the present Universe from forming versus the Universe collapsing into a black holes is $10^{10^{30}}$ to 1.

4. Steven Weinberg, *The First Three Minutes* (New York: Basic Books, 1993).

5. Moché, *Astronomy*, 4th ed., 122.

6. Ronald N. Bracewell, "Communications from superior galactic communities," *Nature* 186(4726): 670–671 (1960). Reprinted in A.G. Cameron, ed., *Interstellar Communication* (New York: W. A. Benjamin, Inc., 1963), 243–248.

7. Clifford Pickover, *The Science of Aliens* (New York: Basic Books, 1999).

8. Hubert Reeves, *Atoms of Silence: An Exploration of Cosmic Evolution* (Cambridge: MIT Press, 1984), 121.

9. "Henrietta Swan Leavitt," http://hoa.aavso.org/posterswan.htm (This web page is part of Hands-On Astrophysics, an AAVSO educational project developed with funds from the National Science Foundation. AAVSO is an acronym for American Association of Variable Star Observers.)

10. Ibid.

11. NASA, "Helium flash," http://imagine.gsfc.nasa.gov/docs/ask_astro/answers/990409a.html

12. J. C. Evans, "Stellar post-main sequence evolution," http://www.physics.gmu.edu/classinfo/astr103/CourseNotes/Text/Lec05/Lec05_pt5_txt_stellarPostMSEvol.htm

13. The carbon cycle is another sequence of thermonuclear reactions that provides much of the energy released by hotter stars. The net result of the carbon cycle is also the fusion of four protons in a helium nucleus.

14. Many particle physicists now think that the so-called "missing neutrinos" may be due to the complicating possibility that there are three or more different types of neutrinos, and the particles can change themselves in a flash from one type to another, thus evading detection. However, to accomplish this, neutrinos must have mass (or weight)—and the Standard Model that physicists have been building up and testing for thirty years suggests that neutrinos have no mass. Recently, a team of Japanese astronomers measured a small mass for the neutrino.

    As we discussed, numerous neutrinos are produced by the proton-proton chain in the Sun. However, neutrinos interact only very weakly with matter. Every second over 100 billion neutrinos from the Sun pass through every square inch of our bodies and virtually none of them interact with us. Because neutrinos interact so weakly with matter, detecting them is very difficult. For example, in the first solar neutrino detection experiment, scientist Ray Davis used 100,000 gallons of cleaning fluid (for the chlorine the fluid contained) in a detector located in a South Dakota gold mine. Davis expected to detect on average of 1.8 solar neutrinos per day. Instead, Davis's observed rate has consistently been much lower than this. Also, the long-term rate, plotted as a function of time, shows an anticorrelation between neutrino rate and sunspot activity.

15. James Kaler, *Stars* (New York: Scientific American Library, 1998), 121–122.

16. Mark Sincell, "A new beginning," *Popular Science*, 258(2): 72 (February 2001).

    Yet another new method for determining cosmic distances involves the study of X-rays from quasars. The quasar's light splinters along several paths as it passes another galaxy along the line of sight to Earth. Slight variations in the time it takes for X-rays to travel a path can reveal the distances to the intervening galaxy. (A single source of light can appear to us as several images when a galaxy acts as a gravitational lens, bending the light rays from the quasar into separate images.) Astronomers hope to use this distance determination method to estimate how rapidly the universe is expanding. For more information on quasar yardsticks, see Robert Irion, "Quixotic quasar may yield cosmic yardstick," *Science* 290(5498): 1885 (December 8, 2000). Also see, R. Cowen, "Variable quasar may help measure the cosmos," *Science News* 158(21): 327 (November 18, 2000).

Chapter 8

1. The information on dimming from white to yellow, red, brown and black comes from the on-line Encyclopedia Britannica (http://www.britannica.com/bcom/eb/article/5/ 0,5716,119405 +10,00.html). However, it appears more common to use the term "red dwarf" when referring to a hydrogen-burning star, not a dimming white dwarf. Similarly, "brown dwarf" usually refers to a low-mass, non-hydrogen burning star and is only rarely used to refer to a dimming white dwarf. Also see, Nick Strobel, "Stages in the life of a star," http:// www.maa.mhn.de/Scholar/ star_evol.html

2. Robert Sawyer, *Calculating God* (New York: Tor Books, 2000), 67.

3. Marcus Chown, *The Magic Furnace* (New York: Oxford University Press, 2001), 169.

4. Sawyer, *Calculating God*, 67. See also note 14 in "Final Thoughts."

   So far, we've been focusing on life's dependence on carbon and not considering that life might be based on other elements. In the past, researchers have speculated that aliens might be based on chains of silicon atoms instead of carbon chains as is the case on Earth. According to chemical laws, there are only two elements capable of forming the long kinds of chain we think are needed for life: carbon and silicon. It appears unlikely that life could exist on Earth that is based on silicon, although a complex system of organic-like chemistry could take place with silicon chains in liquid ammonia instead of water. However, ammonia is liquid only within a narrow range of intensely cold temperatures—making it a less likely environment for life than water. Carbon does have some unique characteristics making it an ideal candidate for life. It can bond to itself in long chains and can form bonds to four other atoms at one time. This allows theoretically for a huge number of different compounds. Note, however, that life could be based on less versatile atoms. For example, it is not necessary for an atom to bond to itself to form long chains. In fact, the chains could be made of two or more atoms in alternation. Professors Gerald Feinberg and Robert Shapiro have speculated that life could be constructed using an alternative chemistry in which the possibilities were not as vast as those of carbon. For example, although the English language can be communicated and stored using 26 letters, it can also be coded as successfully, although less compactly, with 1s and 0s, the binary code used by computers. In the same way, a less complex chemistry could serve as the genetic basis for life, with a large number of components needed in each molecule or cell.

5. The anthropic cosmological principle asserts, in part, that the laws of the Universe are not arbitrary. Instead the laws are constrained by the requirement that they must permit intelligent observers to evolve. Proponents of the anthropic principle note say that human existence is only possible because the constants of physics lie within certain highly restricted ranges. Physicist John Wheeler and others interpret these amazing "coincidences" as proof that human existence somehow determines the design of the Universe. There are alternative explanations of why the Universe appears to be fine-tuned for life, and these explanations do not require a God or designer. For example, our Universe may be one among a huge number of universes. If these universes have random values for their fundamental physical constants, then, just by chance, some of them will permit life to emerge. Using this reasoning, we would be living in one of those special universes, but no designer is needed to set the parameters.

   This area of speculation is controversial. Victor J. Stenger, Emeritus Professor of Physics at the University of Hawaii, has published numerous books and articles that suggest that the conditions for the appearance of a universe with life (and heavy element nucleosynthesis) are not quite as improbable as other physicists have suggested. For more information, see his web site http://spot.colorado.edu/~vstenger. Also see his various books listed at Amazon.com, *The Unconscious Quantum: Metaphysics in Modern Physics and Cosmology* (Amherst, New York: Prometheus, 1995), and his paper, "Cosmythology: is the universe fine-tuned to produce us?" *Skeptic* 4(2): 36–40 (1996). See also note 14 in "Some Final Thoughts."

John Barrow and Frank Tipler, in the *Anthropic Cosmological Principle,* are fascinated by the number of seemingly "coincidental" conditions, events, and physical constants that guide our Universe. For example, they find the number of coincidences involving 10^{39} remarkable.

(Electrical force between a proton and an electron) ÷ (gravitational force between a proton and an electron)	$\dfrac{e^2}{Gm_p m_e}$	$\approx 2.3 \times 10^{39}$
(Age of the Universe time) ÷ (time for light to cross an atom)	$\dfrac{t_u}{e^2 / m_e c^3}$	$\approx 6 \times 10^{39}$
(Planck's constant × speed of light) ÷ (Newton's gravitational constant × squared mass of a proton)	$\dfrac{hc}{Gm_p^2}$	$\approx 10^{39}$
Square root of the number of protons in the observable Universe	$\sqrt{P_u}$	$\approx 10^{39}$

Change these relationships significantly, and our Universe and life as we know it could not exist. For an in-depth discussion on the meaning of these coincidences, see Michael Shermer, *Why People Believe Weird Things* (New York: MJF Books, 1997), 263. Also see Clifford A. Pickover, *The Paradox of God and The Science of Omniscience* (New York: Palgrave, 2002).

6. When I describe white dwarves and neutron stars, I should emphasize that there are separate Chandrasekar limits for the two, and that both kinds of stars are supported by the "exclusion principle"—white dwarves by electron pressure, neutron stars by neutron pressure. Also, there is not an infinite succession of stars composed of ever-smaller subatomic particles, because once the size of the star shrinks below the Schwarzschild limit, absolutely nothing can stand up to the gravitation.

7. The term "Hentriacontane" is a rough translation of the philosopher race's name to English. The races's true name begins with the "H" sound but is rather long and difficult to pronounce. Scientists have therefore adapted the written form "Hentriacontane" when referring to the philosopher race. The name derives from hentriacontane, a hydrocarbon of the paraffin series $CH_3(CH_2)_{29}CH_3$ present in petroleum, many natural waxes, and the pheromones released by the Hentriacontane philosophers. For more information, see Clifford Pickover, *Chaos in Wonderland* (New York: St. Martin's Press, 1995).

8. See Mr. Plex's adventures in Clifford Pickover, *Black Holes: A Traveler's Guide* (New York: Wiley, 1998).

9. Ibid. "Anyone who falls into a black hole will plunge into a tiny central region of infinite density and zero volume . . ." While this statement is true for classical mathematical models of black holes, many researchers believe that when space curvatures reach the value of an inverse Planck length squared, then classical solutions are no longer accurate. In this realm of "quantum gravity" quantum corrections are probably of the same order as the classical curvature. We can only *speculate* as to the exact behavior of bodies taking the cosmic plunge.

10. For more information on the MonkeyGod computer program, see Victor Stenger, *The Unconscious Quantum* (Amherst, New York: Prometheus, 1995). Also see, http://spot.colorado.edu/~vstenger/

11. Chown, *The Magic Furnace*, 169. Chown elegantly writes, "The light elements are fossilized relics from the early Universe and their abundances are directly connected to the extraordinary conditions in the first few minutes of creation. As far as we human beings are concerned, however, it's the heavy elements, not the light elements, which really matter. They, after all, make light possible. And the key to the production of heavy elements are supernova." Note that our Sun is not massive enough to create elements heavier than carbon, oxygen, and nitrogen. In order to make heavier elements, a star has to start out with a mass at least eight times the Sun's.

    Astronomers recognize two different types of supernovae, Type I and Type II. Bright Type I supernovae occur anywhere in a spiral galaxy, but dimmer Type II supernovae are seen only in the spiral arms. Other distinguishing characteristics are described in John Gribbin's book *Stardust* (New Haven, Connecticut: Yale University Press, 2000).

12. Like many red dwarfs, Proxima or Alpha Centauri C is a "flare star." Flare stars can brighten suddenly to many times their normal luminosity. The cause is thought to be a sudden and intense outburst of radiation on or above the star's surface. An increase in radio emission is often detected simultaneously with the optical outburst.

13. Fred Adams and Greg Laughlin, *The Five Ages of the Universe* (New York: Free Press, 1999), xxvi, 75.

14. Ibid. Fred Adams and Greg Laughlin, "A dying universe: The long-term fate and evolution of astrophysical objects." *Review of Modern Physics* 69: 337 (1997). See also Clifford Pickover, *The Science of Aliens* (New York: Basic Books, 1999). Lawrence M. Krauss and Glenn D. Starkman, "The fate of life in the universe," *Scientific American* 281(5): 36–44 (1999).

15. The notion of proton decay is still controversial, but many physicists believe that these extraordinarily long-lived particles eventually die as a result of baryon nonconservation decay paths.

16. Freeman J. Dyson, "Time without end: physics and biology in an open universe." *Reviews of Modern Physics* 51(3): 447–460 (July 1979).

17. The words "vacuum," "nothing," and "void" usually suggest boring, empty space. However, to modern quantum physicists, the vacuum of space has turned out to be rich with complex and unexpected behaviors where a state of minimum energy permits quantum fluctuations. These fluctuations can lead to the temporary formation of particle-antiparticle pairs that usually destroy themselves soon after their creation because there is no source of energy to give the pair permanent existence. These particles are called "virtual particles," and under certain conditions they may separate, become real pairs with positive mass-energy, and become part of the observable world.

18. I discuss brown dwarf priests and related matters in my book, *The Science of Aliens*.

19. Adams and Laughlin, *The Five Ages of the Universe*, 97.

20. George Musser, "The hole shebang," *Scientific American* 283(4) (October 18, 2000).

21. Pickover, *The Science of Aliens*.

22. Timothy Ferris, *The Whole Shebang* (New York: Simon & Schuster, 1997), 308.

23. Jane Robert, *The Seth Material* (Cutchogue, New York: Buccaneer Books, 1995).

24. Anne Rice, *Memnoch the Devil* (New York: Ballantine Books, 1997), 211.

25. John Brooke, "Science and Religion: Lessons from History?" *Science* 282(5396): 1985–1986 (December 11, 1998).

26. Ibid.

27. Adams and Laughlin, *The Five Ages of the Universe*, 84.

28. Ibid., 85.

In 2000, various researchers used supercomputers to simulate collisions of our Galaxy with Andromeda. Although it is very unlikely that any two stars will collide—because of the empty space between stars—our Solar System might get pulled along streamers of stars, 10 thousand light-years long that were gravitationally torn from the galaxies. If we were sucked along for the ride, we might be flung into intergalactic space where the sky would be utterly starless. For more information, see Eric Powell, "Collision Course," *Discover* 21(11): 20 (November, 2000).

Chapter 9

1. Freeman Dyson, "New mercies: the price and promise of human progress." *Science & Spirit* 11(3): 17 (July/August 2000).
2. Stephen Hall, *Mapping the Millennium* (New York: Random House, 1992), 21.
3. Ibid.
4. James Kaler, *Stars* (New York: Scientific American Library, 1998), 13. Interestingly, several constellations are now defunct. For example, you no longer hear of the constellations Antinoüs, Taurus Poniatowski, Noctua the Owl, and Cerberus, although these all have been used for a short time but later rejected.
5. For more on the Bünting map, see for example, Donald P. Ryan, *The Complete Idiot's Guide to Biblical Mysteries* (New York: Alpha Books, 2000), 149; Israel Ministry of Foreign Affairs, "Jerusalem in old maps and views," http://www.israel.org/mfa/go.asp?MFAH00zg0; Derechos Reservados, "Heinrich Bünting," http://www2.ari.net/primavera/1map.html; Jonathan Potter, "Spotlight on antique maps," http://www.jpmaps.co.uk/spotlight.htm; Yale University Library, "A Great Assemblage: Map Collection, Sterling Memorial Library," http://www.library.yale.edu/ exhibition/judaica/mcsml.1.html
6. John Noble Wilford, *The Mapmakers* (New York: Knopf, 2000).
7. Marcus Chown, *The Magic Furnace: The Search for the Origins of Atoms* (New York: Oxford University Press, 2001).
8. Ibid.
9. Ibid.
10. Wendy Freedman, in Robert Lawrence Kuhn, *Closer to Truth* (New York: McGraw-Hill, 2000), 366.
11. Freeman Dyson, "'New Mercies' The price and promise of human progress," *Science & Spirit* 11(3): 16 (July/August, 2000).
12. Freeman J. Dyson, "Time without end: physics and biology in an open universe," *Reviews of Modern Physics* 51(3): 447–460 (July, 1979).
13. Timothy Ferris, *The Whole Shebang* (New York: Simon & Schuster, 1997), 304.
14. Ibid., 305. There is some controversy regarding just how "fine-tuned" these nuclear resonances really are. Steven Weinberg, writing in *The New York Review of Books*, October 21, 1999, suggests that we should be thinking about 3 energy levels:
—— 7.7 MeV (maximum energy of carbon's excited state, beyond which no carbon forms) 0.05 difference from lower level
—— 7.65 MeV (energy of carbon's actual excited state) 0.25 difference from lower level
—— 7.4 MeV (total energy of beryllium 8 nucleus and helium nucleus at rest)

Researchers have suggested that if the carbon resonance energy was higher than 7.7 MeV, no carbon would be formed because the collisions of a helium nucleus and a beryllium 8 nucleus would need a boost of at least 0.3 MeV, a collision energy unlikely to be provided at the temperatures found in stars. So we have several factors to consider. By one criterion, the energy misses being too high by a fractional amount of 0.05 MeV/0.25 MeV, or 20 percent. Weinberg does not find the 20 percent factor a "close call" indicative of "fine-tuning." For more information, see Steven Weinberg, "A Designer Universe?" *The New York Review of Books* XLVI: 46 (October 21, 1999). This article was originally given as a speech at the April 1999 Conference on Cosmic Design of the American Association for the Advancement of Science in Washington, D.C. Also see, M. Livio, D. Holwell, A. Weiss, and J. Truran, "On the Anthropic Significance of the Energy of the O+ Excited State of 12C at 7.644 MeV," *Nature* 340(6231): 281 (July 27, 1989).

Craig Hogan of the University of Washington's Astronomy Department does not find the possible anthropic aspects of the nuclear resonances very compelling. He also notes that getting the production rate of carbon right may not require very precise fine tuning of the resonance because "the structure of stars includes a built-in thermostat that automatically adjusts the temperature to just the value needed to make the reaction go at the correct rate." Nevertheless, Hogan concedes, "It is, however, undeniable and astonishing that starting from a formless hot gas, atoms have developed a sense of their own history." For more details of Hogan's thoughts, see Craig Hogan, "We are Made of Starstuff," *Science* 292(5518): 863 (May 4, 2001). For additional discussion of the anthropic principle and stars, see Clifford A. Pickover, *The Paradox of God and The Science of Omniscience* (New York: Palgrave, 2002).

15. Ibid. Also see, Fred Hoyle "The universe: past and present reflections," *Engineering & Science* (November 1981), 12.
16. Robert Jastrow, "The Astronomer and God," in *The Intellectuals Speak Out about God*, ed. Roy Abraham Varghese (Chicago: Regnery Gateway, 1984), 22.
17. See endnote 5 for chapter 8, which mentions the anthropic cosmological principle and the work of physicist Victor Stenger. Also see, Craig-Pigliucci, "Does God exist?" http://www.leaderu.com/offices/billcraig/docs/craig-pigliucci1.html
18. Stephen W. Hawking, *A Brief History of Time* (New York: Bantam Books, 1988), 123.
19. Fred Adams and Greg Laughlin, *The Five Ages of the Universe* (New York: Free Press, 1999), 198.
20. Ibid., 199.
21. Paul Davies, *Other Worlds* (London: Dent, 1980), 160–161, 168–169.
22. John Barrow and Frank Tipler, *The Anthropic Cosmological Principle* (New York: Oxford University Press, 1986).
23. Paul Davies, *The Mind of God* (New York: Simon & Schuster: 1992), 16.
24. Adams and Laughlin, *The Five Ages of the Universe*, 202–203; Lee Smolin, *Life of the Cosmos* (New York: Oxford University Press: 1997). Roger Penrose and Stephen Hawking have suggested that the expanding Universe is described by the same equations as a collapsing black hole, but with the opposite direction of time. Black holes may be the seeds for other universes. According to John Gribbin, in *Stardust*, the number of baby universes may be proportional to the volume of the parent universe.
25. John Gribbin, *Stardust* (New Haven: Yale University Press, 2000), 222–223.
26. Ibid., 184.
27. Clifford Pickover, *The Science of Aliens* (New York: Basic Books, 1998).
28. Ferris, *The Whole Shebang*, 312.

Stars in the Bible

> *If we wish to understand the nature of the Universe*
> *we have an inner hidden advantage: we are ourselves*
> *little portions of the universe and so carry the answer*
> *within us.*
>
> — Jacques Boivin, "The Single Heart Field Theory"

In the introduction, we spoke of the importance of stars in the Bible, where stars are often signs of God's power and majesty. There is also an immense literature that attempts to come to terms with the Christmas Star represented in Matthew's Gospel. Aside from the Star of Bethlehem, most people don't realize just how pervasive the stars are in the Old and New Testaments.

Obviously, stars are useful for light, navigation, and chronology. They are a source of beauty and inspiration. In the Bible, God knows both the number of stars and the names of stars. The stars are divided into constellations. (Three references to the constellations are found in the Bible, two in Job and the other in Amos.) The Bible condemns astrology but throughout mentions signs in the Heavens. For your interest, here are references to stars in the Bible (mostly the King James Version, but some are in the New International Version).

- Genesis 1:16 And God made two great lights; the greater light to rule the day, and the lesser light to rule the night: he made the stars also.

- Genesis 15:5 And he brought him forth abroad, and said, Look now toward heaven, and tell the stars, if thou be able to number them: and he said unto him, So shall thy seed be.

- Genesis 22:17 That in blessing I will bless thee, and in multiplying I will multiply thy seed as the stars of the heaven, and as the sand which is upon the sea shore; and thy seed shall possess the gate of his enemies . . .

- Genesis 26:4 And I will make thy seed to multiply as the stars of heaven, and will give unto thy seed all these countries; and in thy seed shall all the nations of the earth be blessed . . .

- Genesis 37:9 And he dreamed yet another dream, and told it his brethren, and said, Behold, I have dreamed a dream more; and, behold, the sun and the moon and the eleven stars made obeisance to me.

- Exodus 32:13 Remember Abraham, Isaac, and Israel, thy servants, to whom thou swarest by thine own self, and saidst unto them, I will multiply your seed as the stars of heaven, and all this land that I have spoken of will I give unto your seed, and they shall inherit it for ever.

- Numbers 24:17 I shall see him, but not now: I shall behold him, but not nigh: there shall come a Star out of Jacob, and a Sceptre shall rise out of Israel, and shall smite the corners of Moab, and destroy all the children of Sheth.

- Deuteronomy 1:10 The LORD your God hath multiplied you, and, behold, ye are this day as the stars of heaven for multitude.

- Deuteronomy 4:19 And lest thou lift up thine eyes unto heaven, and when thou seest the sun, and the moon, and the stars, even all the host of heaven, shouldest be driven to worship them, and serve them, which the LORD thy God hath divided unto all nations under the whole heaven.

- Deuteronomy 10:22 Thy fathers went down into Egypt with threescore and ten persons; and now the LORD thy God hath made thee as the stars of heaven for multitude.

- Deuteronomy 28:62 And ye shall be left few in number, whereas ye were as the stars of heaven for multitude; because thou wouldest not obey the voice of the LORD thy God.

- Judges 5:20 They fought from heaven; the stars in their courses fought against Sisera.

- 1 Chronicles 27:23 But David took not the number of them from twenty years old and under: because the LORD had said he would increase Israel like to the stars of the heavens.

- Nehemiah 4:21 So we laboured in the work: and half of them held the spears from the rising of the morning till the stars appeared.

- Nehemiah 9:23 Their children also multipliedst thou as the stars of heaven, and broughtest them into the land, concerning which thou hadst promised to their fathers, that they should go in to possess it.

- Job 3:9 Let the stars of the twilight thereof be dark; let it look for light, but have none; neither let it see the dawning of the day . . .

- Job 9:7 He speaks to the sun and it does not shine; he seals off the light of the stars.

- Job 22:12 Is not God in the height of heaven? and behold the height of the stars, how high they are!

- Job 25:5 Behold even to the moon, and it shineth not; yea, the stars are not pure in his sight.

- Job 38:7 When the morning stars sang together, and all the sons of God shouted for joy?

- Job 38:31–32 Can you bind the beautiful Pleiades? Can you loose the cords of Orion? Can you bring forth the constellations in their seasons or lead out the Bear with its cubs? Do you know the laws of the heavens?

- Psalm 8:3 When I consider thy heavens, the work of thy fingers, the moon and the stars, which thou hast ordained . . .

- Psalm 136:9 The moon and stars to rule by night: for his mercy endureth for ever.

- Psalm 147:4 He telleth the number of the stars; he calleth them all by their names.

- Psalm 148:3 Praise ye him, sun and moon: praise him, all ye stars of light.

- Ecclesiastes 12:1–3 While the sun, or the light, or the moon, or the stars, be not darkened, nor the clouds return after the rain . . .

- Isaiah 13:10 For the stars of heaven and the constellations thereof shall not give their light: the sun shall be darkened in his going forth, and the moon shall not cause her light to shine.

- Isaiah 14:13 For thou hast said in thine heart, I will ascend into heaven, I will exalt my throne above the stars of God: I will sit also upon the mount of the congregation, in the sides of the north . . .

- Isaiah 47:13 Thou art wearied in the multitude of thy counsels. Let now the astrologers, the stargazers, the monthly prognosticators, stand up, and save thee from these things that shall come upon thee.

- Jeremiah 31:35 Thus saith the LORD, which giveth the sun for a light by day, and the ordinances of the moon and of the stars for a light by night, which divideth the sea when the waves thereof roar; The LORD of hosts is his name.

- Ezekiel 32:7 And when I shall put thee out, I will cover the heaven, and make the stars thereof dark; I will cover the sun with a cloud, and the moon shall not give her light.

- Daniel 8:10 And it waxed great, even to the host of heaven; and it cast down some of the host and of the stars to the ground, and stamped upon them.

- Daniel 12:3 And they that be wise shall shine as the brightness of the firmament; and they that turn many to righteousness as the stars for ever and ever.

- Joel 2:10 The earth shall quake before them; the heavens shall tremble: the sun and the moon shall be dark, and the stars shall withdraw their shining . . .

- Joel 3:15 The sun and the moon shall be darkened, and the stars shall withdraw their shining.

- Amos 5:8 Seek him that maketh the seven stars and Orion, and turneth the shadow of death into the morning, and maketh the day dark with night: that calleth for the waters of the sea, and poureth them out upon the face of the earth: The LORD is his name . . .

- Amos 5:26 But ye have borne the tabernacle of your Moloch and Chiun your images, the star of your god, which ye made to yourselves.

- Obadiah 1:4 Though thou exalt thyself as the eagle, and though thou set thy nest among the stars, thence will I bring thee down, saith the LORD.

- Nahum 3:16 Thou hast multiplied thy merchants above the stars of heaven: the cankerworm spoileth, and fleeth away.

- Matthew 2:2 Saying, Where is he that is born King of the Jews? for we have seen his star in the east, and are come to worship him.

- Matthew 2:7 Then Herod, when he had privily called the wise men, enquired of them diligently what time the star appeared.

- Matthew 2:9 When they had heard the king, they departed; and, lo, the star, which they saw in the east, went before them, till it came and stood over where the young child was.

- Matthew 2:10 When they saw the star, they rejoiced with exceeding great joy.

- Matthew 24:29 Immediately after the tribulation of those days shall the sun be darkened, and the moon shall not give her light, and the stars shall fall from heaven, and the powers of the heavens shall be shaken.

- Mark 13:25 And the stars of heaven shall fall, and the powers that are in heaven shall be shaken.

- Luke 21:25 And there shall be signs in the sun, and in the moon, and in the stars; and upon the earth distress of nations, with perplexity; the sea and the waves roaring . . .

- Acts 7:43 Yea, ye took up the tabernacle of Moloch, and the star of your god Remphan, figures which ye made to worship them: and I will carry you away beyond Babylon.

- Acts 27:20 And when neither sun nor stars in many days appeared, and no small tempest lay on us, all hope that we should be saved was then taken away.

- 1 Corinthians 15:41 There is one glory of the sun, and another glory of the moon, and another glory of the stars: for one star differeth from another star in glory.

- Hebrews 11:12 Therefore sprang there even of one, and him as good as dead, so many as the stars of the sky in multitude, and as the sand which is by the sea shore innumerable.

- 2 Peter 1:19 We have also a more sure word of prophecy; whereunto ye do well that ye take heed, as unto a light that shineth in a dark place, until the day dawn, and the day star arise in your hearts . . .

- Jude 1:13 Raging waves of the sea, foaming out their own shame; wandering stars, to whom is reserved the blackness of darkness for ever.

- Revelation 1:16 And he had in his right hand seven stars: and out of his mouth went a sharp two-edged sword: and his countenance was as the sun shineth in his strength.

- Revelation 1:20 The mystery of the seven stars which thou sawest in my right hand, and the seven golden candlesticks. The seven stars are the angels of the seven churches: and the seven candlesticks which thou sawest are the seven churches.

- Revelation 2:1 Unto the angel of the church of Ephesus write; These things saith he that holdeth the seven stars in his right hand, who walketh in the midst of the seven golden candlesticks . . .

- Revelation 2:28 And I will give him the morning star.

- Revelation 3:1 And unto the angel of the church in Sardis write; These things saith he that hath the seven Spirits of God, and the seven stars; I know thy works, that thou hast a name that thou livest, and art dead.

- Revelation 6:13 And the stars of heaven fell unto the earth, even as a fig tree casteth her untimely figs, when she is shaken of a mighty wind.

- Revelation 8:10 And the third angel sounded, and there fell a great star from heaven, burning as it were a lamp, and it fell upon the third part of the rivers, and upon the fountains of waters . . .

- Revelation 8:11 And the name of the star is called Wormwood: and the third part of the waters became wormwood; and many men died of the waters, because they were made bitter.

- Revelation 8:12 And the fourth angel sounded, and the third part of the sun was smitten, and the third part of the moon, and the third part of the stars; so as the third part of them was darkened, and the day shone not for a third part of it, and the night likewise.

- Revelation 9:1 And the fifth angel sounded, and I saw a star fall from heaven unto the earth: and to him was given the key of the bottomless pit.

- Revelation 12:1 And there appeared a great wonder in heaven; a woman clothed with the sun, and the moon under her feet, and upon her head a crown of twelve stars . . .

- Revelation 12:4 And his tail drew the third part of the stars of heaven, and did cast them to the earth: and the dragon stood before the woman which was ready to be delivered, for to devour her child as soon as it was born.

- Revelation 22:16 I, Jesus, have sent mine angel to testify unto you these things in the churches. I am the root and the offspring of David, and the bright and morning star.

☆ APPENDIX 2

Updates and Breakthroughs

He showed me a little thing, the quantity of
a hazelnut, in the palm of my hand, and it
was round as a ball. I looked thereupon with
eye of my understanding and thought: What
may this be? And it was answered generally
thus: It is all that is made.

— Julian of Norwich, 14th Century

While writing *The Stars of Heaven*, I have monitored current scientific breakthroughs and theories relating to stellar processes. Here is a sampling of recent information on stars. Most information comes from news articles that have come across my desk during the writing of this book.

Welcome to the Age of Megastars

Researchers have recently theorized that the first stars that evolved in our universe were behemoths, several hundreds of times the mass of the Sun. Currently the biggest stars in the cosmos are about a hundred times as massive as the Sun. (Recall in chapter 5 we spoke of some huge stars like Betelgeuse and Mu Cephei.) Eta Carina in the Milky Way is about 120 times more massive than the Sun, and the most massive star known, HDE 269801 in the Large Magellanic Cloud, is about 190 times the mass of the Sun. Theorists once thought that stars more massive than 100 Suns would be unstable, because the greater a star's mass, the hotter it is and the greater the pressure from internal radiation. However, the first stars in the universe were born when only hydrogen and helium existed. For a variety of theoretical reasons, the heavy elements available in stars today make them less stable and unable to grow much bigger than 100 solar masses. For example, today's molecules of carbon monoxide and hydrogen cyanide can cool down intersteller gasses.

A typical megasun in the early universe probably had a surface temperature of 100,000 K and shined as brightly as 10 million Suns. If such a megastar was placed at the location of the our nearest star in the night sky, Alpha Centauri, the megastar would appear 50 times as bright as the full Moon. These megastars would live about 3 million years and then violently explode. For example, a 250-solar-mass star would be blown apart in a "hypernova" and leave no remnant behind. During the hypernova, the megastar would burn more brilliantly than 100 galaxies and leave the newly formed heavy elements scattered to stupendous distances. Fifty solar masses of iron alone would be spewed into space. For a 300-solar-mass star, the hypernova would produce more energy than 10 billion galaxies and then leave behind a 30-solar-mass black hole. About 100 grams of your own blood have been forged in the megastars present in the universe's infancy. [For more information, see Marcus Chown, "Titanic," *New Scientist* (September 11, 2000) http://www.newscientist.com/nlh/0909/titanic.html]

The supermassive stars thought to populate the early universe are known as population III stars and may be the progenitors for population II stars, the old red stars found in galaxies today, which are rich in heavy elements. [For more information, see Linda Rowan, "Old metals, new stars," *Science* 290(5489): 13 (October 6, 2000).]

The term "hypernova" or "supernova" has also been used to describe the blast of energy released when any supermassive star collapses into a black hole. Recently, astronomers have suggested that a massive star first explodes as a supernova, shedding its iron into space and leaving behind a spinning neutron star. Eventually the neutron stars slows down and then implodes with a burst of gamma rays to form a black hole. According to the new theory, the gamma ray burst lights up the older supernova shell, which would explain observations of a dense, iron-rich material already millions of kilometers from the center of the explosion by the time the burst takes place. [For more information see, Govert Schilling, "Gamma ray bursts may pack a one-two punch," *Science* 290(5493):926–927 (November 3, 2000).]

Neutron Star Twists Einstein

Like stirring a thick syrup with a rotating spoon, Einstein's general theory of relativity predicts that heavy objects in the universe should twirl the fabric of spacetime around themselves. Astronomers may finally have found evidence for this effect by studying the rapid variations in brightness of X-rays emitted by neutron stars. Gas ripped from nearby stars orbits in a spinning accretion disk around a neutron star at speeds close to the speed of light. As the gas spirals inward, the gas heats up and emits X-rays. Astronomers have measured oscillations in X-ray brightness from the neutron stars and suggest that the specific frequencies observed may represent the difference in rotation rate between the orbiting gas and the spinning neutron star. The precise nature of the oscillations seems to suggest a frame dragging (spatial warping) effect on the space near the neutron star. Frame dragging can cause the accretion disk to wobble like a Frisbee. [For more information, see Govert Schilling, "Neutron stars imply relativity's a drag," *Science* 289(5484): 1448 (September 1, 2000); Ron Cowen, "Neutron star twists Einstein's theory," *Science News* 158(6): 150 (2000).]

Speaking of neutron stars, Stephan Rosswog and colleagues from the University of Leicester, in England, and the University of Basel, in Switzerland recently suggested that most of the platinum, gold, and other heavy elements on Earth was manufactured during huge explosions of colliding neutron stars, hundreds of millions of years before the solar system formed. In April, 2001, Stephan Rosswog said, "It's exciting to think that the gold in wedding rings was formed far away by colliding stars." Traditionally, astrophysicists believed that elements, such as oxygen and carbon, are created when dying stars explode into supernovas, but researchers have also been puzzled by data that suggests these stellar explosions do not produce enough heavy elements to account for heavy element abundance on Earth. The new theories of heavy element production rely on supercomputer simulations of neutron star collisions that eventually produce a black hole. The heat produced during the collisions is around a billion degrees Celsius, allowing the necessary nuclear reactions to take place.

Some astrophysicists believe the r-process mechanism for producing these heavy elements also occurs in supernovae. (R-process reactions are rapid reactions, probably occurring inside supernovae, in which heavy elements are formed as atomic nuclei capture neutrons. This is in contrast to s-process reactions, the slower reactions in giant stars in which heavy elements are created as atomic nuclei capture neutrons.) Supporters of the supernova theory of heavy element production argue that collapses of binary neutron stars happen too infrequently to produce all the gold on Earth. For further information, see the popular treatment in: Chris Fontaine, "Exploding Energy of Neutron Stars May have Spun Gold—Scientists Say Precious Metals Landed on Earth," Associated Press (reprinted by many newspapers and on the Internet web), April 8, 2001.

Brown Dwarfs on Parade

Throughout this book we've discussed brown dwarfs. These objects are too massive to be planets yet too small to be stars. These dwarfs probably arise the same way stars do, from the gravitational collapse of giant gas clouds, but these dwarfs are too small to sustain fusion.

Researchers in 1999 were using the Chandra X-ray Observatory to "stare" at a brown dwarf— an old, failed star—for many hours when it suddenly sprang to life! Brown dwarfs, as they grow older, lose their hot outer atmosphere, or corona, and can't readily generate X-ray flares. They lack the mass to burn steadily. On this occasion, a brown dwarf known as LP 944-20 emitted an X-ray flare. This suggests that although the older brown dwarfs have no corona, they do have magnetic fields that can sometimes generate storms beneath the surface that lead to X-ray flares. [For more information, see Ron Cowen, "X-ray flare from a dim source," *Science News* 158(5): 72 (July 2, 2000).]

In 2000, researchers conducted a deep infrared photometric survey in the Orion nebula when they found that 30 percent of their 500 infrared sources were brown dwarfs. [For more information, see Linda Rowan, "Free-floating planets in Orion," *Science* 288(5467): 773 (May 5, 2000).]

The coolest stars with just enough mass to fuse hydrogen are the *M-dwarfs* (see chapter 3). Two new classes of brown dwarfs have been added to the cool end of the stellar spectrum. The *L-dwarfs* (1,300 to 2,000 Kelvin) are slightly cooler and less massive than M-dwarfs. *T-dwarfs* are cooler and lighter than the L-dwarfs. Both of these new dwarfs cannot sustain hydrogen fusion. Researchers have recently discovered hundreds of T-dwarfs and tens of L-dwarfs. Even the cool T-dwarfs may have magnetic fields that create occasional stellar flares. [For more information, see Linda Rowan, "Cooler dwarf stars," *Science* 289(5480): 697 (August 4, 2000).]

In 2000, scientists used the Hubble Space Telescope's near-infrared camera to study these objects that are usually too dim to be seen in visible light. In particular, the cameras were turned to study brown dwarfs in a cluster of Milky Way stars known as IC 348. Because the cluster is young, the brown dwarfs are still somewhat "bright," making them more apparent to Hubble's cameras. Researchers note that brown dwarfs are cool enough to contain water in their atmosphere. They're 10 to 80 times the mass of Jupiter, and although common, they are not so plentiful as to account for most of the unseen matter that astronomers believe inhabit a giant halo around our galaxy. [For more information, see Ron Cowen, "Taking a census of brown dwarfs," *Science News* 158(11): 168 (September 9, 2000).]

Some astronomers believe that the growing population of extrasolar planets may be misleading. Nearly half of the so-called planets recently "discovered" orbiting around other stars may actually be brown dwarfs. The standard method for detecting extrasolar planets can only detect the minimum mass of an orbiting object. The actual mass may be much greater. [For more information, see Ron Cowen, "Are most extrasolar planets hefty impostors?" *Science News* 158(18): 227 (October 28, 2000).]

The Lithium Connection

Today, astronomers often analyze the spectra of faint objects in the sky when hunting for brown dwarfs. All stars burn lithium in their cores when a proton collides with the isotope lithium 7, which then splits into two helium atoms. The reaction might be symbolized by a diagram in which black circles are neutrons and unfilled circles are protons:

$$^{7}\text{Li} + \text{Proton} \Rightarrow {^{4}\text{He}} + {^{4}\text{He}}$$

●○●○●○● + ○ ⇨ ●○●○ + ●○●○

In contrast to stars, all but the most massive brown dwarfs cannot achieve the core temperatures necessary for lithium consumption, so brown dwarfs usually retain the element. This means that scientists can locate brown dwarfs by analyzing their spectra for the presence of lithium. Normal stars quickly consume whatever lithium they might have originally had. Interestingly, because brown dwarfs do not consume much hydrogen, in the distant future, when all stars have burned out and consumed their hydrogen through fusion, brown dwarfs will be the main vaults of hydrogen in the universe. [For more information, see Gibor Basri, "The Discovery of Brown Dwarfs," *Scientific American* 282(4): 77–79 (April 2000).]

Magnetars

In chapter 8, we discussed pulsars, rapidly spinning collapsed stars. Astronomers can see each rotation as a regular pulse of radiation. Recently, there were new additions to the pulsar family. At only 700 years of age, PSR J1846-0258 in the Kes 75 supernova remnant appears to be a *magnetar*, a pulsar with a relatively slow rotation period and especially intense magnetic field that is more than 100 trillion times that of Earth's. The precise nature of the difference between pulsars and magnetars has yet to be determined. [See R. Bennett, "Young pulsar has a split personality." *Science News* 158(8):117 (August 19, 2000).]

Black Holes

Astronomers studying the supermassive black hole at the center of our Galaxy discovered it belching bubbles as large as our entire solar system. The bubbles are formed when stars and gas fall into the black hole. As the material swirls around the black hole, creating an accretion disk, the material heats up due to friction. Eventually, it can become so hot that it bubbles away into cooler regions of space. The bubbles coming out of the accretion disk create tiny blips in the black hole's radio signal. [For more information, see Michael Mayer, "Galactic geysers," *Popular Science* 257(4): 26 (October, 2000).]

Traditionally, astronomers have estimated the mass of a black hole residing at the center of a galaxy by clocking the motion of *individual* stars in the surrounding galaxy. In 2000, astronomers Karl Gebhardt of the University of Texas and Laura Ferrarese of Rutgers University discovered that the mass of the black hole is proportional to the *overall* motion of the stars in the galaxy's central bulge, which is relatively easy to measure. Another fairly new way of estimating black hole mass involves studying the light from quasars—the tremendous energy beacons coming from many galactic black holes. Using these approaches, the researchers determined that many of the galactic black holes are continually gobbling surrounding gas and stars. It had been previously thought that the supermassive black holes finished growing soon after their host galaxies formed. [For more information, see Mark Sincell, "Ravenous black holes never say diet," *Science* 291(5501): 28 (January 5, 2001). Also see, Ron Cowen, "X rays unveil secret lives of black holes," *Science News* 159(1): 6 (January 6, 2001).]

Mirror Stars and Supernovae

Scientists postulate the existence of an entire universe of "shadow matter," made of particles similar to neutrons, protons, and electrons. According to this theory, the shadow universe shares our space, but it is difficult to detect because mirror photons are invisible. However, for a variety of theoretical reasons, mirror matter may affect our universe by exerting gravitational attraction. Scientist suggest that this mirror matter may help explain the mystery of cosmic gravitation and in particular help explain why ordinary matter coalesces so easily to produce huge clusters of visible galaxies. Mirror material could build the cosmic skeleton upon which our galaxies form.

Mirror matter also plays a role in the missing solar neutrino problem discussed in this book. Part of the reason solar neutrino emission is only half that expected may be that certain neutrinos transform into mirror neutrinos that are undetectable. Astronomers and theoreticians are also currently searching for mirror stars. If a mirror star explodes as a supernova, there would be a burst of mirror neutrinos, which may transform into detectable (normal) neutrinos on our "side" of the universe. Although we can never see a mirror supernova itself, we would see a strange burst of neutrinos far from any visible star in our universe. Similarly, mirror supernovas might create pairs of mirror electrons and mirror positrons that would convert into ordinary electrons and positrons that annihilate themselves in our universe. This annihilation could be seen as a big burst of visible light. [For further information, see Ron Cowen, "Through the looking glass," *Science News* 158(11): 173–175 (September 9, 2000). Also see, Marcus Chown, "Shadow worlds," June 17, 2000, http://www.newscientist.com/features/ features.jsp?id=ns224324]

Other theorists are considering the wild possibility that when a star collapses and then explodes as a supernova, the high temperatures can boil off gravitons into dimensions beyond ordinary three spatial dimensions. (Gravitons are hypothetical particles that transmit gravity according to quantum theory.) These researchers suggest that the visible universe could lie on a membrane "floating" within a higher dimensional space. The extra dimensions help unify the forces of nature and could contain parallel universes that may exist on their own membranes less than a millimeter away from ours. "Dark matter" is explained by ordinary stars and galaxies on nearby sheets. The gravity of this dark matter affects us by taking shortcuts through the extra dimensions. [For more information see, Nima Arkani-Hamed, Dimopoulos Savas, and Georgi Dvali, "The Universe's unseen dimensions," *Scientific American* 283(2): 62–69 (August 2000).]

Cosmic Carbon Bomb

In 1999, using NASA's Rossi X-ray Timing Explorer satellite, astronomers watched as carbon on the surface of a neutron star detonated in a three-hour thermonuclear explosion. This was the first known cosmic explosion fueled solely by carbon rather than hydrogen or helium.

The strange carbon blast was produced by binary star 4U 1820-30, which consists of a dwarf star orbiting a neutron star. Gas from the dwarf flows in a spiral pattern around the neutron star. When some of the dwarf's gas collides with the neutron star's surface, a compressed slurry of hydrogen and helium is formed. Pressures and temperatures can get sufficiently high in the slurry layer that the elements flash-fuse in a thermonuclear explosion. Each blast leaves carbon, one of the byproducts of helium fusion. Gradually a layer of carbon several hundred meters thick reaches a critical temperature and ignites a carbon bomb that rages for hours. [For more information, see Robert Irion, "Astronomers spot their first carbon bomb," *Science* 290(5495): 1279 (November 17, 2000).]

Water's Role in Making Stars

The water molecule plays a key role during the early stages of star formation. Water is an important source of oxygen and contributes to the cooling of the circumstellar gas, which helps to remove the excess energy produced during the collapse of a protostar. When stars begin to form, water may be present either as gas or as ice on the surface of dust particles. When the star forms from the gravitational collapse of dust and gas clouds, the cloud is normally quite cold (about 20 Kelvin), and almost all the water is in the icy coats of dust grains. In 2000, the Infrared Space Observatory (ISO) confirmed the presence of a large amount of water in protostars. [For further information, see Brunella Nisini, "Water's role in making stars," Science 290(5496): 1513 (November 24, 2000).]

Galactic Cannibalism

Various astronomical models suggest that billions of years ago the Milky Way had dozens of small companion galaxies. Now there are only 11 such companions, the closest galaxy being the Sagittarius Dwarf galaxy discovered a few years ago on the opposite side of the Milky Way from the Sun. This galaxy appears to have been stretched, torn and assimilated by our galaxy. Even now, the Milky Way is being invaded. In the next 100 million years, the Sagittarius Dwarf galaxy will move though the disk of our own Milky Way galaxy yet again. We won't be in danger, but large gravitational tidal forces might pull the Dwarf apart. The Dwarf may have survived many prior encounters because it contains a great deal of low-density dark matter that holds it together gravitationally during its encounters with the Milky Way. [For further information , see "Galactic Archeology," Scientific American 284(6): 21–22 (June 2001). See also, Robert Nemiroff (MTU) and Jerry Bonnell (USRA), "Sagittarius Dwarf to Collide with Milky Way," http://antwrp.gsfc.nasa.gov/apod/ap980216.html]

How to Save the Earth

In this book, we discussed how in just a few billion years, the hydrogen fuel in our Sun will be exhausted in its core, and the Sun will begin to die and dramatically expand, becoming a red giant. At some point, Earth's oceans will boil away and all life will perish. Recently, a team of scientists led by Donald Korycansky of the University of California at Santa Cruz has developed a visionary plan that would add another six billion years to our lives. In the March 2001 Astrophysics and Space Science, Korycansky shows how Earth's orbit can be increased if a suitable asteroid can be made to fly in front of Earth. In doing so, the asteroid imparts some of its orbital energy to Earth and shifts Earth to a slightly larger orbit. Korycansky and colleagues determined that for Earth to have the same sunlight levels as it does today, our planet would have to be nudged outward once every 6,000 years by a 10^{16} metric ton object. Although this method seems far fetched, it actually uses technology that is only a few decades away. One drawback is that our Moon would probably be stripped away during all the cosmic nudging, and the Moon plays an important role in regulating Earth's climate. [For further information, see Mark A. Garlick, "Save the Earth," Scientific American 284(6): 24D (June 2001). Also see, Donald Korycansky, Gregory Laughlin, and Fred C. Adams, "Astronomical Engineering: a Strategy for Modifying Planetary Orbits," Astrophysics and Space Science 275: 349–366 (2001); Donald Korycansky, "Astroengineering," http://www.ucolick.org/~kory/]

The Paradox of the Sun's Hot Corona

Like a boiling pot of chocolate perched on a cold linoleum floor, the sun's hot outer layers sit on a colder surface. As we discussed in chapter 6, the sun's temperature decreases steadily as one moves outward from its core (15 million degrees K) to the photosphere (5,800 degrees K), but then the temperature gradient reverses as one enters the chromosphere (15,000 degrees K) and the corona (2 million degrees K). Considering that heat is presumably generated beneath the photosphere, why would the corona get so hot? Astronomers suggest that magnetic fields cause the anomalous coronal heating. Magnetic fields loop through the solar atmosphere and interior to form an intricate web of magnetic structures. Magnetic fields are the strongest in the corona and can transport energy in many forms away from the sun. Scientists still discuss the possibility that the energy for coronal heating derives from the magnetic field itself in addition to the field acting as a passive conduit for energy from the subphotosphere. Various methods for studying of corona heating using spacecraft and fast imaging of the corona during eclipses are described in: Bhola N. Dwivedi and Kenneth J. H. Phillips, "The Paradox of the Sun's Hot Corona," *Scientific American* 284(6): 41–47 (June 2001). See also, David H. Hathaway, "3D Coronal Magnetic Fields," http://science.nasa.gov/ssl/pad/solar/ 3dfields.htm.

Richard Feynman and Stars

One of my favorite quotations about stars makes a fitting conclusion to this "Updates and Breakthroughs" Appendix. Theoretical physicist and Nobel Prize winner Richard Feynman (1918–1988) wrote in *The Feynman Lectures on Physics*:

"The stars are made of the same atoms as the earth." I usually pick one small topic like this to give a lecture on. Poets say science takes away from the beauty of the stars—mere gobs of gas atoms. Nothing is "mere." I too can see the stars on a desert night, and feel them. But do I see less or more? The vastness of the heavens stretches my imagination—stuck on this carousel my little eye can catch one-million-year-old light. A vast pattern—of which I am a part—perhaps my stuff was belched from some forgotten star, as one is belching there. Or see them with the greater eye of Palomar, rushing all apart from some common starting point when they were perhaps all together. What is the pattern, or the meaning, or the why? It does not do harm to the mystery to know a little about it. For far more marvelous is the truth than any artists of the past imagined! Why do the poets of the present not speak of it? What men are poets who can speak of Jupiter if he were like a man, but if he is an immense spinning sphere of methane and ammonia must be silent? [Richard Phillips Feynman, Matthew L. Sands, Robert B. Leighton, *The Feynman Lectures on Physics: Commemorative Issue*, vol. 1 (New York: Addison-Wesley, 1994).]

Further Reading

The universe: a device contrived for the perpetual astonishment of astronomers.

— Arthur C. Clarke

See "Notes" section for additional books, web sites, and articles.

Books

Fred Adams and Greg Laughlin, *The Five Ages of the Universe* (New York: Free Press, 1999).

Marcus Chown, *The Magic Furnace: The Search for the Origins of Atoms* (New York: Oxford University Press, 2001).

Timothy Ferris, *The Whole Shebang* (New York: Simon & Schuster, 1997).

John Gribbin with Mary Gribbin, *Stardust* (New Haven, Connecticut: Yale University Press, 2000).

James Kaler, *Stars* (New York: Scientific American Library, 1992).

Rudolf Kippenhahn, *100 Billion Suns* (Princeton: Princeton University Press, 1983).

Dina Moché, *Astronomy*, 4th ed. (New York: John Wiley & Sons, 1998).

Clifford Pickover, *Chaos in Wonderland: Visual Adventures in a Fractal World* (New York: St. Martin's Press, 1995). Among other things, this book provides amazingly simple computer recipes that allows readers to build their own globular clusters. Globular clusters are spherical groups of stars, containing thousands or even millions of stars. Astronomers have found more than 200 globular clusters in the Milky Way Galaxy. In my research, I simulated and drew dozens of such clusters and used computer graphics programs to fly around them in real time. Hundreds of globular clusters surround the disk of our Galaxy, concentrating near its center. Some astrophysicists think the centers of globular clusters contain small black holes. In *Chaos in Wonderland*, you can become the master of your own pocket universe of stars.

Martin Rees, *Just Six Numbers* (New York: Basic Books, 2000).

Hubert Reeves, *Atoms of Silence* (Cambridge: MIT Press, 1984).

Robert J. Sawyer, *Calculating God* (New York: Tor Books, 2000).

Iosif Shklovskii, *Stars: Their Birth, Life, and Death* (New York: W. H. Freemman, 1978).

Web Sites

As many readers are aware, Internet Web sites come and go. Sometimes they change addresses or completely disappear. The Web site addresses listed here provided valuable background information when this book was written. You can, of course, find numerous other

web sites relating to stars and astronomy by using Web search tools such as the ones provided at www.google.com.

Bill Baity, "Stars,"
    http://exobio.ucsd.edu/Astronomy/stars.htm
Chris Dolan, "The Constellations and their Stars,"
    http://www.astro.wisc.edu/~dolan/constellations/
Carol Gerten-Jackson, "Vincent van Gogh, The Starry Night,"
    http://cgfa.kelloggcreek.com/gogh/p-gogh41.htm
Riccardo Giovanelli and Martha Haynes, "The evolution of the sun"
    http://astrosun.tn.cornell.edu/courses/astro201/evol_sun.htm
Calvin J. Hamilton, "The sun,"
    http://www.solarviews.com/eng/sun.htm
Gordon D. Holman, "High Energy Solar Spectroscopic Imager (HESSI) home page,"
    http://hesperia.gsfc.nasa.gov/hessi/index.html
Gordon D. Holman, "Solar flare,"
    http://hesperia.gsfc.nasa.gov/sftheory/fulldisk.htm
Gordon D. Holman, "Solar prominence,"
    http://hesperia.gsfc.nasa.gov/hessi/images/eit_19990514_0718_304.jpg
Gordon D. Holman, "High Energy Solar Spectroscopic Imager (HESSI)"
    http://hesperia.gsfc.nasa.gov/hessi/index.html
Gordon D. Holman, "HESSI,"
    http://hesperia.gsfc.nasa.gov/hessi/images/ppthessi4.jpg
Lake County Astronomical Society, "Parallax and Schlessinger,"
    http://yang.interaccess.com/~purcellm/articles/yerkdist.htm
Robert Nemiroff (MTU) and Jerry Bonnell (USRA), "Astronomy picture of the day,"
    http://antwrp.gsfc.nasa.gov/apod/
Robert Nemiroff (MTU) and Jerry Bonnell (USRA), "TRACE and the active sun"
    http://antwrp.gsfc.nasa.gov/apod/ap980515.html
NASA Newsroom, "NASA photo gallery,"
    http://www.nasa.gov/gallery/photo/index.html
Office of Public Outreach, Space Telescope Science Institute, "Hubble Space Telescope public
    pictures," http://oposite.stsci.edu/pubinfo/Pictures.html
J. J. O'Connor and E. F. Robertson, "Friedrich Wilhelm Bessel,"
    http://www-groups.dcs.st-and.ac.uk/~history/Mathematicians/Bessel.html
Sten Odenwald, "Ask the Astronomer," http://itss.raytheon.com/cafe/qadir/astarss.html
Richard W. Pogge, "The once and future sun,"
    http://www-astronomy.mps.ohio-state.edu/~pogge/Lectures/vistas97.html
Stanford University, "Stanford Solar Center," http://solar-center.stanford.edu/
Steve Suess (NASA), "The interstellar probe and the heliopause,"
    http://science.nasa.gov/ssl/pad/solar/suess/Interstellar_Probe/ISP-Intro.html
Joseph S. Tenn, "The Bruce Medalists—Schlesssinger,"
    http://www.phys-astro.sonoma.edu/BruceMedalists/BM2S.html

About the Author

$\mathbf{C}$LIFFORD A. PICKOVER received his Ph.D. from Yale University's Department of Molecular Biophysics and Biochemistry. He graduated first in his class from Franklin and Marshall College, after completing the four-year undergraduate program in three years. His many books have been translated into Italian, Greek, German, Japanese, Chinese, Korean, Portuguese, French, Italian, and Polish. He is author of the popular books: *The Paradox of God and the Science of Omniscience* (Palgrave/St. Martin's Press, 2002), *The Zen of Magic Squares, Circles, and Stars* (Princeton University Press, 2001), *Dreaming the Future* (Prometheus, 2001), *Wonders of Numbers* (Oxford University Press, 2000), *Surfing Through Hyperspace* (Oxford University Press, 1999), *The Science of Aliens* (Basic Books, 1998), *Time: A Traveler's Guide* (Oxford University Press, 1998), *Strange Brains and Genius: The Secret Lives of Eccentric Scientists and Madmen* (Plenum, 1998), *The Alien IQ Test* (Basic Books, 1997), *The Loom of God* (Plenum, 1997), *Black Holes - A Traveler's Guide* (Wiley, 1996), and *Keys to Infinity* (Wiley, 1995). He is also author of numerous other highly-acclaimed books including *Chaos in Wonderland: Visual Adventures in a Fractal World* (1994), *Mazes for the Mind: Computers and the Unexpected:* (1992), *Computers and the Imagination* (1991) and *Computers, Pattern, Chaos, and Beauty* (1990), all published by St. Martin's Press—as well as the author of over 200 articles concerning topics in science, art, and mathematics. He is also coauthor, with Piers Anthony, of *Spider Legs*, a science-fiction novel once listed as Barnes and Noble's second best-selling science-fiction title.

Pickover is currently an associate editor for the scientific journals *Computers and Graphics* and *Theta Mathematics Journal,* and is an editorial board member for *Odyssey, Idealistic Studies, Leonardo,* and *YLEM.* He has been a guest editor for several scientific journals.

Editor of the books *Chaos and Fractals: A Computer Graphical Journey* (Elsevier, 1998), *The Pattern Book: Fractals, Art, and Nature* (World Scientific, 1995), *Visions of the Future: Art, Technology, and Computing in the Next Century* (St. Martin's Press, 1993), *Future Health* (St. Martin's Press, 1995), *Fractal Horizons* (St. Martin's Press, 1996), and *Visualizing Biological Information* (World Scientific, 1995), and coeditor of the books *Spiral Symmetry* (World Scientific, 1992) and *Frontiers in Scientific Visualization* (Wiley, 1994), Dr. Pickover's primary interest is finding new ways to continually expand creativity by melding art, science, mathematics, and other seemingly-disparate areas of human endeavor.

The *Los Angeles Times* recently proclaimed, "Pickover has published nearly a book a year in which he stretches the limits of computers, art and thought." Pickover received first prize in the Institute of Physics' "Beauty of Physics Photographic Competition." His computer graphics have been featured on the cover of many popular magazines, and his research has recently received considerable attention by the press—including *CNN's* "Science and Technology Week," *The Discovery Channel, Science News, The Washington Post, Wired,* and *The Christian Science Monitor*—and also in international exhibitions and museums. *OMNI* magazine recently described him as "Van Leeuwenhoek's twentieth century equivalent." *Scientific American* several times featured his graphic work, calling it "strange and beautiful, stunningly realistic." *Wired* magazine wrote, "Bucky Fuller thought big, Arthur C. Clarke thinks big, but Cliff Pickover outdoes them both." Among his many patents, Pickover has received U.S. Patent 5,095,302 for a 3-D computer mouse, 5,564,004 for strange computer icons, and 5,682,486 for black-hole transporter interfaces to computers.

Dr. Pickover is currently a Research Staff Member at the IBM T. J. Watson Research Center, where he has received 30 invention achievement awards, three research division awards, and four external honor awards. For many years, Dr. Pickover was the Brain-Boggler columnist for *Discover* magazine. He currently is the Brain-Strain columnist for *Odyssey*, and he has also published popular puzzle calendars.

Dr. Pickover's hobbies include the practice of Ch'ang-Shih Tai-Chi Ch'uan and Shaolin Kung Fu, raising golden and green severums (large Amazonian fish), and piano playing (mostly jazz). He is also a member of The SETI League, a group of signal processing enthusiasts who systematically search the sky for intelligent, extraterrestrial life. Visit his web site, www.pickover.com, which has received over 500,000 visits. He can be reached at P.O. Box 549, Millwood, New York 10546-0549 USA.

Index

Absorption lines, 21–22
Active galactic nucleus, 183
Adams, Fred, 174
AGN, 183
Algol, 83
Aliens, 33–34, 37
Alien senses, 37
Alpha capture, 152
Alpha Centauri, 12
Alpha particle, 152
Andromeda, 86, 186, 210
Antares, 19
Anthropic principle, 158, 193, 205,
     207–208
Aristotle, 7
Asimov, Isaac, 192
Atoms, 20–29
AU, 32

Baby universes, 195
Baha'u'llah, vi
Balmer series, 23–26, 29, 38
Barrow, John, 194
Baxter, Stephen, 203
Bernard's star, 20
Beryllium-8, 151–153
Bessel, Freidrich, 10–11, 87
Betelgeuse, 78–79, 88, 124–125, 134
Bethlehem, xv–xvi, 199
Bible, xv, 8–10, 212–214
Big Bang, 57, 86–87, 119
Big Dipper, 49
Binary stars, 81–84, 114, 218, 221
Black dwarfs, 144, 147, 173

Black holes, 15, 159, 166, 181, 220
Blackbody, 36
Blue giants, 75, 124
Blue stragglers, 33–35
Blueshift, 46–48
Bohr, Niels, 20, 25, 28, 38
Boime, Albert, xv
Boltzmann, Ludwig, 36
Bracewell, Ronald, 133
Brahe, Tycho, xvii
Brightness, 60–65, 89, 117, 130
Brown dwarfs, 70, 77, 173–174, 178
Bünting, Heinrich, 189

Canis Major, 79, 82, 87
Cannon, Annie, 40, 54–55
Carbon, 158, 192, 221
     bomb, 221
     burning, 160, 192
     resonance, 154–157, 192, 210–211
Cash, Johnny, vi
Castor, 81–84
Cepheids, 129, 137
Chandrasekar limit, 163
Chown, Marcus, vii
Chromosphere, 93–116
Coincidence, 193, 205, 207–208
Color, 19
Constellations, 189
Contact, 30
Convective zone, 94
Copernicus, 7
Core, 83–116, 160–161
Corona, 93–116, 223

Corona Borealis, 48
Coronal mass ejection, 108–110
Cosmic microwave background
    radiation, 72, 86–87
Cygnus, 10

Davies, Paul, 194
Death of Universe, 172–177
Density of stars, 80, 94
Deuterium, 140
Dipper, Big, 49, 81–82
Distance modulus, 65, 130
Dog Star, 87
Doppler shift, 30, 46–48, 57, 70
Double stars, 81–84, 114, 218, 221
Drake, Frank, 13, 32
Dupree, Andrea, 78
Dwarfs, 70, 74–75, 80, 88–89, 114, 124,
    136, 144–148, 160, 163, 172–174,
    178, 219
  Brown, 70, 77, 173–174, 178, 219
  Yellow, 173
Dyson sphere, 32–33
Dyson, Freeman, xx, 32–33, 177, 192

Eagle Nebula, 122
Egyptians, 87–88
Einstein, Albert, 166
Electromagnetic radiation, 17–22, 36, 77,
    86, 106
Elements
  generation of, 30, 87, 114, 128, 142–
    144, 153, 160–162, 191, 218
  in stars, 30, 41–42, 55, 73, 93, 114,
    128, 136, 142–144, 153, 160–162,
    191, 218
Emission lines, 21–22
End of the World, xx
Energy levels, 20–24
Ethnocentric geography, 49
Evolution of stars, 80, 117–141
Extraterrestrials, 33–34, 37

Ferris, Timothy, 185
Feynman, Richard, 223

Flares, solar, 101, 106–108
Fraunhofer, Joseph von, 21
Fusion, helium, 126–127, 148, 151
Fusion, 34, 89, 103, 114, 121, 123–126,
    139, 148, 151, 161

Gaiman, Neil, x, 57
Galaxies, 53, 57, 73, 86, 114–115, 186,
    188, 210, 222
Galilei, Galileo, xvii, 7–10, 188
Geocentric theory, 7, 188
Giants, 54, 73–75, 78–80, 88, 114, 128,
    146, 157, 217
God, 70, 147, 169, 185–186, 193, 207
Gold, 161
Goodricke, John, 130
Granulation, solar, 96–98
Gravity, 121

Hassan, Ihab, x
Hawking, Stephen, 193
Heliacal rising, 88
Heliocentrism 7, 188
Heliopause, 105–106
Heliosheath, 105
Heliosphere, 106
Helium, 30, 34, 87, 89, 93, 103, 123–126,
    139–140, 152, 194
  flash, 126, 138
  fusion, 126–127, 148
Hemicentrism, 49
Hertzsprung-Russell, 74, 80, 88–89, 122
Hipparchus, xvii
Hipparcos, 5
Hogan, Craig, 211
Hoyle, Fred, 154, 192
Hubble Space Telescope, 92
Hydrogen, 30, 34, 87, 93, 103, 114, 123–
    124, 139, 152
Hydrostatic equilibrium, 120

Ice Age, 115
Integers, 38
Ionization, 23, 45, 55
Islam, viii
Isotopes, 33

Janssen, Pierre, 30
Jastrow, Robertk, 193
Jerusalem map, 189–190
Joshua, 8–9

Kant, Immanuel, 92
Kepler, Johannes, xvii, 68

Laughlin, Gregory, 174
Leavitt, Henrietta, 137–138
Lewis, C. S., 13
Lifetime of stars, 34, 41, 68, 80, 88, 103
Light emission, 20–24
Light-year (defined), 5
Limb (solar), 95
Lithium, 219
Local Group, 53, 57, 186
Luminosity, 60–65, 68, 74, 77, 80, 89,
    117, 127, 130, 138
  method, 12
Lyman series, 23–26

Magnesium, 160
Magnetars, 220
Magnetic fields, 100–101, 107, 220, 223
Magnetograph, 101
Magnetosphere, 106
Magnitude, 62–63, 130
Main sequence, 75, 80, 86, 88, 123–124,
    148
Maps, 190
Mass of stars, 76, 173
Mass-luminosity relation, 76, 130, 138
Maunder minimum, 102
Maxwell, James, 17
Megastars, 88, 217
Messengers, 132–133
Milky Way, 15, 53, 73, 186, 188
Mira variables, 129–131
Mirfak, 53
Mirror stars, 221
Mizar, 81–82
Moche, Dina, vii
Modulus, 65, 130
Motion, stars, 46–48, 53

Mu Cephei, 88
Multiple Universes, 188–190, 194

Nebulae, 42, 92, 116, 120, 122, 143–146
    planetary, 42, 92, 116, 136, 143–146
Nebular theory, 92, 116
Neon, 160
Nephilim, 117–118
Neutrinos, 204, 206, 221
Neutron star, 159, 163, 184, 218
Novas, 73, 114, 121, 147, 158–161, 221
Nucleosynthesis, 87, 114, 128, 142–144,
    153, 160–162, 191, 218

Olbers' paradox, 68–70
Orion, 79, 87, 89, 120, 199
Oxygen, 157

Paradox, Olbers', 68–70
Parallax, 1–14, 65
Parsec (defined), 5
Paschen series, 23–26
Pauli exclusion principle, 162
Payne-Gaposhkin, Cecilia, 40–41, 54–55
Perignon, Dom, 1
Pfund series, 26
Photosphere, 22 93–116
Photosynthesis, 91
Piano, electromagnetic, 18
Planetary nebulae, 42, 92, 116, 136,
    143–146
Plants, 89
Pleiades, 199
Poe, Edgar, Allan, 70
Population I and II stars, 73
Power 60–61
Prominences, 108
Proplyds, 92, 116, 120
Proton-proton reaction, 139
Protoplanetary disks, 89, 116, 120
Protostars, 120–122, 148
Proxima Centauri, 12
Ptolemy, xvii
Pulsars, 164, 180

Quantum mechanics, 168, 185

Radiation
    background, 72, 86–87
    electromagnetic, 17–22, 36, 77, 86,
        106
Radiative zone, 94, 98
Red dwarfs, 74–75, 78, 88, 124, 135, 172,
    173
Red giants, 114, 124, 128, 134, 146, 157,
    222
Red shift method, 12
Redshift, 46–48
Reese, Martin, 194
Reeves, Hubert, 33–34
Resonance, 154–157, 210–211
R-process, 218
RR Lyarae, 129–131
Rydberg-Ritz, 27–29, 38

S-process, 218
Sagittarius B2, 54
Sagittarius Dwarf galaxy, 222
Sawyer, Robert, 185
Schlesinger, Frank, 11–12
Schwarzchild radius, 166
Shadow matter, 221
Shklovsky, Samuilovich, 33
Shock waves, 114
Sirius, 1, 82, 87
Smilodon, x
Solar
    flares, 101, 106–108
    nebular theory, 92
    power, 61, 89
    prominences, 108–109
    system, 92–93, 114
    wind, 105
Spectra 17–29, 40–41, 53–54, 63
Spectroheliograph, 104
Starry Night, xiii–xv
Stars
    Bethlehem, xv–xvi, 199
    binary, 81–84
    black hole, 15, 159, 166, 181, 220
    brightness of, 60–65, 74–75, 89, 117

    classes of, 41–42, 53–54, 63,65, 73–75,
        88, 134–135
    color, 19
    death of, 34, 177
    density of, 80, 94
    distances, 1–14, 140, 206
    dwarfs, 74–75, 80, 88–89, 114, 124,
        136, 145–148, 160, 163, 172, 173–
        174, 178, 219
    elements in 30, 41–42, 55, 73–75, 128,
        136, 142–144, 153, 160–162, 191,
        218
    evolution of, 80, 117–141
    generation of, 114
    giants, 54, 73–75, 78, 88, 114, 128,
        146, 157, 217
    in art, xiii–xv
    in Bible, xv, 212–214
    in Islam, xvii
    lifetime of, 68, 80, 88, 103
    luminosity of, 60–65, 68, 74–75, 77,
        80, 89, 117, 130, 138
    mass of, 76, 173
    mirror, 221
    motion of, 46–47, 53
    neutron, 159, 163, 184
    number of, xvii, 70
    organic chemicals in, 54
    power of, 60–61, 89
    pulsars, 164
    rejuvenation of, 34
    size of, 78–80, 88–89
    spectra, 16–29, 40–41, 57, 65
    temperature, 16–19, 36, 46, 53, 65, 77,
        95, 115, 144
    variable, 129, 148
    water in, 221
    Wolf-Rayet, 42–44
Stefan-Boltzman Radiation Law, 77
Steinbeck, John, 38
Stenger, Victor, 207
Strahl, 105
Sun, 45, 75, 78, 80, 89, 93–116, 114–115,
    121, 124, 134
    cycle, 102

death of, 34, 135
flares, 101, 106–108
prominences
rotation, 103
wind 105
Sunspots, 99–103, 108, 115
Supergiants, 74, 78–80
Supernovas, xii, 114, 73, 121, 140, 147, 158–161, 221
Swedish Vacuum Solar Telescope, 97

Technetium, 32–33
Temperature, 16–19, 36, 40, 46, 53, 65, 74–75, 77, 80, 95, 115, 144
Tiger, x
Tipler, Frank, 194
Triple alpha process, 153, 158
Tucanae, 34–35
Tull, Jethro, 13
TX Camelopardalis, 131

UFOs, 13
Universe
death of, 172–177

expansion of, 57, 87
multiple, 188–190, 194
size of, xviii, 70

van Gogh, Vincent, xiii
Variable stars, 129, 148
Vega, 19, 30, 68
Verschuur, Gerrit, 12

Water in stars, 222
White dwarfs, 74, 80, 89, 114, 136, 145–148, 160, 163, 173
White giants, 78
Wien, Wilhelm, 36
Wind, solar, 105
Wolf-Rayet stars, 42–44
Women in astronomy, 54–55
Woodstock, xiii

Yellow dwarfs, 173

Zeta Geminorum, 130

7.6549, 155